Lecture Notes in Control and Information Sciences

Edited by M. Thoma and A. Wyner

163

V. L. Mehrmann (Ed.)

The Autonomous Linear Quadratic Control Problem

Theory and Numerical Solution

Springer-Verlag
Berlin Heidelberg New York
London Paris Tokyo
Hong Kong Barcelona Budapest

Editor
Volker Ludwig Mehrmann
Fakultät für Mathematik
Universität Bielefeld
Postfach 86 40
4800 Bielefeld 1
Germany

ISBN 3-540-54170-5 Springer-Verlag Berlin Heidelberg New York
ISBN 0-387-54170-5 Springer-Verlag New York Berlin Heidelberg

Printing: Mercedes-Druck, Berlin
Binding: B. Helm, Berlin
61/3020-543210   Printed on acid-free paper.

For Lennart

Preface: In this book a survey is given on the state of the art in theory and the numerical solution of general linear quadratic optimal control problems with differential algebraic equation constraints. This field is very active in the last ten years and in particular the step from standard ordinary differential equation constraints to differential algebraic equation constraints has received a lot of interest. Although the theory for the standard case is known since the sixties the development of numerical methods is not complete. For the descriptor case this is even more the case. Here not even the development of the theory can be considered near to complete. On the other hand these problems occur in many applications, in particular in engineering and it is therefore important to have a solid mathematical theory as well as reliable numerical algorithms. For this reason we review and extend theory and numerical methods for these problems together and we hope that this book helps increase the cooperation between pure mathematicians, numerical analysts and practioners.

The basis of this book was laid in my Habilitation thesis at the University of Bielefeld in 1987/1988. Apart from the survey of the current literature, it contains many new results and methods that were obtained with co-authors in the last 6 years. Without the joint efforts of these people this work would have never been finished. I wish to express my thanks to these co-authors, Gregory Ammar, Angelika Bunse-Gerstner, Ralph Byers, Ton Geerts, Gerd Krause, Peter Kunkel, Nancy Nichols and David Watkins and also to my students Engkan Tan, Ulrike Flaschka and Dagmar Zywietz and last but not least to Ludwig Elsner, for many helpful discussions.

I further wish to thank Regine Hollmann and Marion Matz for help in typesetting this book.

Contents

§ 1 Introduction.

We study the following continuous and discrete time, autonomous, linear quadratic optimal control problom.

(1.1) Continuous time problem
Minimize

$$S(y(t), u(t)) = \frac{1}{2}(y(T)^* M\ y(T)$$

$$+ \int_{t_0}^{T} [y(t)^* Q\ y(t) + u(t)^* R\ u(t) + u(t)^* S^*\ y(t) + y(t)^* S\ u(t)] dt)$$

subject to the dynamics

$$E\dot{x}(t) = A\ x(t) + B\ u(t),\ t_0 < t \le T$$
$$x(t_0) = x^0$$
$$y(t) = C\ x(t),\ t_0 \le t \le T,$$

where $A, E \in \mathbf{C}^{n,n}$, $B \in \mathbf{C}^{n,m}$, $S \in \mathbf{C}^{p,m}$, $C \in \mathbf{C}^{p,n}$, $M = M^*$, $Q = Q^* \in \mathbf{C}^{p,p}$, $R = R^* \in \mathbf{C}^{m,m}$, $y(t) \in \mathbf{C}^p$, $x^0, x(t) \in \mathbf{C}^n$, $u(t) \in \mathbf{C}^m$, $t_0, t, T \in \mathbf{R}$. A, B, C, E, M, Q, R, S are constant in time.

(By Q^* we denote the conjugate transpose of Q and $\dot{x} := \frac{d}{dt}x(t)$.)

(1.2) Discrete time problem
Minimize

$$S'(y, u) = \frac{1}{2}(y_K^* M\ y_K$$

$$+ \sum_{k=k_0}^{K-1} [y_k^* Q\ y_k + u_k^* R\ u_k + u_k^* S^*\ y_k + y_k^* S\ u_k])$$

subject to the difference equation

$$E\ x_{k+1} = A\ x_k + B\ u_k, \quad k = k_0, k_0 + 1, \ldots, K - 1$$
$$x_{k_0} = x^0$$
$$y_k = C\ x_k,\ k = k_0, \ldots, K,$$

where now $x = (x_{k_0}, \ldots, x_K)$, $u = (u_{k_0}, \ldots, u_K)$, $y = (y_{k_0}, \ldots, y_K)$ and $x^0, A, B, C, E, M, Q, R, S$ are as in (1.1).

In our considerations we include the case $K, T = \infty$. Note that we have formulated everything in the complex version. Almost all of the following results hold similary in the real case. In those cases, where the real cases are different we will point that out.

A simple but very instructive example of (1.1) is the following control problem. It is a simple mathematical model of a DC–motor. The example is discussed in detail in Knobloch/Kwakernaak [K 18].

1.3 EXAMPLE.   *(Control of DC–motor)*

$$u \longrightarrow \quad AMPLIFIER \quad \rightarrow \quad DC-MOTOR \quad \longrightarrow \quad \theta.$$

A DC–motor is controlled by an input voltage u via an amplifier. The output of the system is the rotation angle θ of the motor axis. Let ω be the angular velocity of the motor axis then the state of the system is $x = \begin{bmatrix} \theta \\ \omega \end{bmatrix}$. We have the following linearized equations of motion

(1.4)
$$\dot{\theta} = \omega$$
$$j\dot{\omega} = -b\omega + ku,$$

or in matrix form

(1.5)
$$\begin{bmatrix} 1 & 0 \\ 0 & j \end{bmatrix} \dot{x} = \begin{bmatrix} 0 & 1 \\ 0 & -b \end{bmatrix} x + \begin{bmatrix} 0 \\ k \end{bmatrix} u, \quad x(0) = 0,$$

where

— j is the moment of inertia of the rotor,
— b is the friction coefficient,
— k is the amplification proportionality factor.

The output of the system is clearly

(1.6)
$$y = [1 \ 0]x = \theta.$$

If we consider the cost functional

(1.7)
$$S(y,u) = \int_0^T \frac{1}{2}[y(t)^T y(t) + \rho u(t)^2]dt$$

then we have the following system matrices

(1.8)
$$E = \begin{bmatrix} 1 & 0 \\ 0 & j \end{bmatrix}, \ A = \begin{bmatrix} 0 & 1 \\ 0 & -b \end{bmatrix}, \ B = \begin{bmatrix} 0 \\ k \end{bmatrix}, \ C = [1 \ 0],$$
$$Q = 1, \ R = \rho, \ S = 0.$$

We shall come back to Example 1.3 later.

The number of applications of problem (1.1), (1.2) is immense and spread over almost all sciences. It would be impossible to list all possible references. A wide spread of different applications, with different physical properties of the system is given in the following list of references, which is far from being representative or complete.

Anderson/Moore [A 6], [A 7], Anderson/Vongpanitlerd [A 8], Athans [A 14], Athans/Falb [A 15], Bar–Ness [B 2], Bittanti/Bolzern [B 16], Brocket [B 18], Burns [B 38], Casti [C 5], Dorato/Levis [D 13], Emami–Naeini [E 5], Emami–Naeini/Franklin [E 6], [E 7], Führer [F 6], Hautus [H 5], Hughes/Skelton [H 13], Jacobson [J 2], Kailath [K 3], Kalman [K

4], [K 5], Kalman/Bertram [K 6], Kalman et al [K 7], [K 8], Knobloch/Kwaakernaak [K 18], Kwakernaak/Sivan [K 28], Leontief [L 18], Levine/Athans [L 19], Müller [M 18], Ramirez/Park [R 1], de Spinadel [S 12], Trentelmann [T 4], Udilov [U 1], Willems [W 12], [W 13], Wonham [W 18], [W 19] and many other books and papers concerned with optimal control, systems theory etc.

Problems (1.1), (1.2) are two of the simplest and best understood optimal control problems. They are often used as linear models for nonlinear problems (as unfortunately most "real world" problems are).

The typical approach to solve optimization problems with constraints is to use Lagrange multiplier theory or the Pontryagin maximum principle [P 15]. In Section 3 we will apply this principle to the two problems and show that one obtains analogous results for the discrete and continuous case. Standard references for these approaches are Athans/Falb [A 15], Brocket [B 14], Casti [C 4], Dorato/Levis [D 13], von Escherich [E 8], Kailath [K 3], Kalman [K 4], Knobloch/Kwakernaak [K 18], Kwakernaak/Sivan [K 28], Sage [S 1], Willems [W 11] to name a few.

For the case $E = I$, the application of this theory to problems (1.1), (1.2) is well–known. But recently, initated by problems with second order differential equation constraints, e.g. Brenan et al [B 17], Campbell [C 1], [C 2], Dorf [D 14], Hughes/Skelton [H 13], or Leontief models in economics, e.g. Leontief [L 18], Luenberger/Arbel [L 23], Rosenbrock [R 11], Tan [T 1], there has been great interest in systems with singular E, known as descriptor systems, semistate systems, generalized state space systems or just singular systems. For references on this different kind of singular systems, for example see Bunse–Gerstner et al [B 32], [B 33], Bender/Laub [B 11], [B 12], Campbell [C 1], [C 3], Chapman et al [C 6], Cobb [C 9], [C 10], De Souza et al [D 8], Geerts [G 7], [G 8], [G 9], [G 10], Geerts/Mehrmann [G 11], Lewis/Ozcaldiran [L 20], Mehrmann [M 8], [M 9], [M 10], Pandolfi [P 4], [P 5], Shayman [S 4], [S 5], Shayman/Zhou [S 6], [S 7], Thorp [T 3], Van der Weiden/Bosgra [V 1], Verghese et al [V 11], [V 12], Willems et al [W 14], Yip/Sincovec [Y 1].

In Section 3 we discuss the application of the maximum principle to problems (1.1), (1.2) in the general case, i.e. including E singular. In both the continuous and discrete case, application of the maximum principle yields linear boundary value problems

(1.9)
$$\mathcal{A} \, \dot{z}(t) = \mathcal{B} \, z(t)$$
$$f(z(t_0)) = 0, \; g(z(T)) = 0,$$

in the continuous case and

(1.10)
$$\mathcal{A}' \, z_{k+1} = \mathcal{B}' \, z_k$$
$$f'(z_{k_0}) = 0, \; g'(z_K) = 0,$$

in the discrete case, where the matrix pencils $\alpha \, \mathcal{A} - \beta \, \mathcal{B}$, $\alpha \, \mathcal{A}' - \beta \, \mathcal{B}'$ have the forms

(1.11)
$$\alpha \mathcal{A} - \beta \mathcal{B} = \alpha \begin{bmatrix} A & 0 & B \\ C^*QC & A^* & C^*S \\ S^*C & B^* & R \end{bmatrix} - \beta \begin{bmatrix} E & 0 & 0 \\ 0 & -E^* & 0 \\ 0 & 0 & 0 \end{bmatrix}$$

in the continuous case and

(1.12)
$$\alpha \mathcal{A}' - \beta \mathcal{B}' = \alpha \begin{bmatrix} A & 0 & B \\ C^*QC & -E^* & C^*S \\ S^*C & 0 & R \end{bmatrix} - \beta \begin{bmatrix} E & 0 & 0 \\ 0 & -A^* & 0 \\ 0 & -B^* & 0 \end{bmatrix}$$

in the discrete case, and the components are as before. The analysis of the boundary value problems can be reduced to the study of spectral properties of the matrix pencils $\alpha\,A - \beta\,B$, $\alpha\,A' - \beta\,B'$. In Section 4 we discuss this topic.

EXAMPLE 1.3 CONTINUED.

If we apply the maximum principle to Example 1.3, we obtain that the optimal solution to the control problem given by (1.5), (1.6), (1.7) exists if and only there exists a costate function $\mu(t) = \begin{bmatrix} \mu_1(t) \\ \mu_2(t) \end{bmatrix}$ such that $x(t) = \begin{bmatrix} x_1(t) \\ x_2(t) \end{bmatrix}$, $\mu(t)$, $u(t)$ satisfy the boundary value problem

(1.15)
$$\begin{bmatrix} 0 & 1 & 0 & 0 & 0 \\ 0 & -b & 0 & 0 & k \\ 1 & 0 & 0 & 0 & 0 \\ 0 & 0 & 1 & -b & 0 \\ 0 & 0 & 0 & k & \rho \end{bmatrix} \begin{bmatrix} x_1(t) \\ x_2(t) \\ \mu_1(t) \\ \mu_2(t) \\ u(t) \end{bmatrix} = \begin{bmatrix} 1 & 0 & 0 & 0 & 0 \\ 0 & j & 0 & 0 & 0 \\ 0 & 0 & -1 & 0 & 0 \\ 0 & 0 & 0 & -j & 0 \\ 0 & 0 & 0 & 0 & 0 \end{bmatrix} \begin{bmatrix} \dot{x}_1 \\ \dot{x}_2 \\ \dot{\mu}_1 \\ \dot{\mu}_2 \\ \dot{u} \end{bmatrix}$$
$$\begin{bmatrix} x_1(0) \\ x_2(0) \end{bmatrix} = \begin{bmatrix} 0 \\ 0 \end{bmatrix}, \quad \begin{bmatrix} 1 & 0 \\ 0 & j \end{bmatrix} \begin{bmatrix} \mu_1(T) \\ \mu_2(T) \end{bmatrix} = \begin{bmatrix} 0 \\ 0 \end{bmatrix}.$$

The study of differential equations of the form (1.9), (1.10) is well–known and goes back to Kronecker [K 21], e.g. Gantmacher [G 5], recent studies of this topic are given in Campbell [C 2], [C 3] or Brenan et al [B 17].

Using the properties of the matrix pencils $\alpha\,A - \beta\,B$, $\alpha\,A' - \beta\,B'$ we are able to discuss uniqueness of solutions of (1.1), (1.2), in particular feedback solutions, i.e. controls

(1.16)
$$u(t) = T\,y(t),$$

or

(1.17)
$$u_k = T\,y_k,$$

which are of primary interest in applications. In particular we are interested in feedback solutions, which are not only optimal but also asymptotically stabilizing, i.e. they lead to closed loop systems with $\lim_{t\to\infty} x(t) = 0$ and $\lim_{k\to\infty} x_k = 0$.

In Section 5 we shall show analogously to the well–known cases, that these optimal, stabilizing feedback solutions exist, under generalizations of the usual system theoretic assumptions of stabilizability and detectability.

The optimal stabilizing feedback solutions can be obtained via the solution of algebraic Riccati equations or Riccati differential equations.

The mathematical properties of these equations (typically for $E = I$, $C = I$, $S = 0$) are studied for example in Bucy [B 19], [B 20], [B 21], Casti [C 4], Coppel [C 11], Gaalman [G 1], Gohberg et al [G 13], [G 14], Hautus [H 5], Incertis [I 1], Kalman [K 4], Kučera [K 22], [K 23], [K 24], [K 25], Lancaster et al [L 1], [L 2], [L 3], [L 4], Martensson [M 4], Mehrmann [M 8], [M 10], Molinari [M 14], [M 15], [M 16], [M 17], Packer/Bullock [P 1], Patel/Toda

[P 8], Payne/Silverman [P 9], Potter [P 16], Ran [R 2], Ran/Rodman [R 3], Silverman [S 8], Simaan [S 9], Singer/Hammarling [S 10], Vaughan [V 8], [V 9], Willems [W 11], [W 12], [W 13], Wimmer [W 15], [W 16], [W 17] to name only a few.

A recent review of the properties of algebraic Riccati equation is given in Lancaster/Rodman [L 4]. In Section 6 we will extend some of the results on Riccati equations and their relationship to optimal control problems to the general case $E \neq I$. Algebraic Riccati equations also arise in other types of control problems and thus can also be discussed as an independent topic, e.g. Petersen [P 11].

We discuss conditions under which the optimal controls for problems (1.1), (1.2) are of the form

$$(1.18) \qquad u(t) = -R^{-1}(S^*C + B^*X(t)E)x(t)$$

in the continuous case with $T < \infty$, where $X(t)$ satisfies the Riccati differential equation

$$(1.19) \qquad \begin{aligned} -E^* \dot{X}(t)E &= A^*X(t)E + E^*X(t)A \\ &\quad - (B^*X(t)E + S^*C)^*R^{-1}(B^*X(t)E + S^*C) + C^*QC, \\ E^*X(T)E &= M \end{aligned}$$

or if $T = \infty$, $X(t)$ is constant and is the unique Hermitian positive semidefinite solution of the algebraic Riccati equation

$$(1.20) \qquad 0 = A^*XE + E^*XA - (B^*XE + S^*C)^*R^{-1}(B^*XE + S^*C) + C^*QC.$$

In the discrete case we discuss conditions such that for $K < \infty$

$$(1.21) \qquad u_k = -(R + B^*X_kB)^{-1}(AX_kB + C^*S)^*x_k$$

where X_k satisfies the Riccati difference equation

$$(1.22) \qquad \begin{aligned} E^*X_kE &= -E^*X_{k+1}E + A^*X_{k+1}A \\ &\quad + C^*QC - (B^*X_{k+1}A + S^*C)^*(R + B^*X_{k+1}B)^{-1}(B^*X_{k+1}A + S^*C), \\ E^*X_KE &= M \end{aligned}$$

or if $K = \infty$, then X_k is constant, and the unique Hermitian positive semidefinite solution of the discrete algebraic Riccati equation

$$(1.23) \quad 0 = -E^*XE + A^*XA + C^*QC - (B^*XA + S^*C)^*(R + B^*XB)^{-1}(B^*XA + S^*C)$$

EXAMPLE 1.3 CONTINUED.

In Example 1.3 with $T = \infty$ we then obtain the algebraic Riccati equation

$$(1.24) \qquad \begin{aligned} 0 &= \begin{bmatrix} 0 & 0 \\ 1 & -b \end{bmatrix}\begin{bmatrix} x_{11} & x_{12} \\ x_{21} & x_{22} \end{bmatrix}\begin{bmatrix} 1 & 0 \\ 0 & j \end{bmatrix} + \begin{bmatrix} 1 & 0 \\ 0 & j \end{bmatrix}\begin{bmatrix} x_{11} & x_{12} \\ x_{21} & x_{22} \end{bmatrix}\begin{bmatrix} 0 & 1 \\ 0 & -b \end{bmatrix} \\ &\quad - \left([0\ k]\begin{bmatrix} x_{11} & x_{12} \\ x_{21} & x_{22} \end{bmatrix}\begin{bmatrix} 1 & 0 \\ 0 & j \end{bmatrix}\right)^* \rho\left([0,k]\begin{bmatrix} x_{11} & x_{12} \\ x_{21} & x_{22} \end{bmatrix}\begin{bmatrix} 1 & 0 \\ 0 & j \end{bmatrix}\right) + \begin{bmatrix} 1 & 0 \\ 0 & 0 \end{bmatrix} \end{aligned}$$

and the optimal control is

$$(1.25) \qquad u(t) = -\frac{1}{\rho} \, [0 \ k] \begin{bmatrix} x_{11} & x_{12} \\ x_{21} & x_{22} \end{bmatrix} \begin{bmatrix} 1 & 0 \\ 0 & j \end{bmatrix} \begin{bmatrix} x_1(t) \\ x_2(t) \end{bmatrix},$$

where $\begin{bmatrix} x_{11} & x_{12} \\ x_{21} & x_{22} \end{bmatrix}$ is the unique Hermitian positive semidefinite solution of (1.24).

Due to the large number of applications of problem (1.1), (1.2) in real world engineering, there is demand for good, reliable and easy to use numerical algorithms that can be applied to solve these problems. As we have seen there are several different ways to solve (1.1), (1.2), which also lead to several different numerical methods. We discuss most of the known methods, put them in the same framework and analyse their numerical properties.

In Section 7 we dicuss a number of canonical forms, Schur forms, Hessenberg forms which simplify the analysis and construction of numerical methods but also may work as tools for the theoretical analysis, for which we, however, use different canonical forms.

The basis for an error analysis of the discussed methods is the perturbation analysis of control problems (1.1), (1.2). This is still a topic of very active research at the moment, but we give a brief survey of the current state in Section 8.

Some of the numerical methods that we discuss only work for special cases, and often more general problems can be reduced to these special cases. There are essentially two compression methods that we discuss in Section 9, one due to Bunse–Gerstner et al [B 33], [B 34], [B 35], essentially dealing with the case E singular and one due to Van Dooren [V 4] treating the case that R is singular.

Most of the numerical methods for problems (1.1), (1.2) are based on the solution of the algebraic Riccati equations (1.18), (1.20). Unfortunately numerical methods usually give inaccurate results due to roundoff errors. Some of the methods for solving Riccati equations are not even numerically stable, thus if they are applied nevertheless, there should be a possibility to improve the solution for example by iterative refinement.
In Section 10 we discuss a general defect correction method due to Mehrmann/Tan [M 10], which as we show, can be combined with any other method for the numerical solution of the Riccati equation to work as a procedure for iterative refinement.

A particular example of such a defect correction method is Newton's method or variations of it applied to (1.18), (1.20). Newton's method for this problem is the topic of many recent publications, e.g. Allwright [A 1], Anderson [A 5], Arnold [A 9], Hammarling [H 3], Hewer [H 6], Kenney/Laub [K 12], Kenney et al [K 14], Kleinman [K 17], Sandell [S 2].
We discuss Newton's method in detail in Section 11.

Another historically early approach was to compute solutions of (1.1), (1.2) by computing particular eigenvalues and eigenvectors of the corresponding matrix pencils

$$\alpha A - \beta B, \ \alpha A' - \beta B'$$

given by (1.11), (1.12). See Fath [F 1], Gaalman [G 1], MacFarlane [M 1], Martensson [M 4], Michelson [M 12], Potter [P 16], Vaughan [V 10].

One may use standard numerical techniques for the computation of eigenvalues and eigenvectors of the corresponding matrices or matrix pencils in numerical linear algebra to solve this problem.

It was observed by Martensson [M 4] that it is not really necessary to compute all eigenvalues and eigenvectors of the pencils (1.11), (1.12) to obtain solutions. It is sufficient to compute certain deflating subspaces of (1.11), (1.12). Based on this idea, new methods were proposed, one of which is the sign function method which we discuss in Section 12.

This approach has been widely used recently, in particular since, it has a potential to be parallelized or vectorized, i.e. to be used on modern supercomputers. See Balzer [B 1], Barraud [B 3], [B 4], Beavers/Denham [B 6], [B 7], [B 8], [B 9], Byers [B 39], [B 43], Denman/Beavers [D 7], Gardiner/Laub [G 6], Howland [H 10], Kenney/Laub [K 11], Kenney et al [K 13], Roberts [R 6] in this context.

The most popular approaches today are nevertheless methods that are based on the computation of Schur like forms of the matrix pencils $\alpha\, A - \beta\, B$, $\alpha\, A' - \beta\, B'$. Well-known methods like the QR- and QZ-algorithm can be applied to compute these forms. Arnold/Laub [A 9], [A 10], Emami–Naeini [E 5], Laub [L 7], [L 9], [L 13], Lee [L 17], Maki [M 2], Michelson [M 12], Pappas et al [P 5], Van Dooren [V 4], Walker et al [W 1] and others. All these methods use elementary unitary transformations to transform the pencils to a Schur like form. These transformations are described in Section 13 and then applied in the general Schur approach discussed in Section 14. Unfortunately the general QR- or QZ-algorithm do not make any use of the nice algebraic properties of the pencils $\alpha\, A - \beta\, B$, $\alpha\, A' - \beta\, B'$ discussed in Section 4. A framework of a method that would make use of these properties is also discussed in Section 14.

The key idea for such a method was proposed by Paige/Van Loan [P 2] which then led to a number of other publications which discuss the use of the structure of $\alpha\, A - \beta\, B$, $\alpha\, A' - \beta\, B'$ in numerical procedures. Bunse–Gerstner [B 25], [B 26], [B 27], Bunse–Gerstner/Mehrmann [B 28], [B 29], Bunse–Gerstner et al [B 30], [B 32], [B 36], Byers [B 39], [B 41], [B 42], Byers/Mehrmann [B 46], Clements/Glover [C 8], Flaschka et al [F 2] Mehrmann [M 6], [M 8]. We describe these methods in Sections 15, 16.

The basis for the exploitation of the special structure in these problems, is the use of symplectic geometry, which is hidden in these problems. Typical references dealing with this questions are: Ammar [A 2], Ammar/Martin [A 3], Artin [A 13], Belifante/Kolman [B 10], Burgoyne/Cushman [B 37], Ciampi [C 7], Elsner [E 1], [E 2], [E 4], Hua [H 11], [H 12], Laub/Meyer [L 14], for the theoretical background and Bunse–Gerstner [B 25], [B 26], Byers [B 39], Della–Dora [D 1], [D 2], Elsner [E 2], [E 4], Mehrmann [M 8], Paige/Van Loan [P 2] for theoretical results concerning the numerical analysis. None of these methods for the solution of algebraic Riccati equations, however, is completely satisfactory, since the optimal method would be one which

— exploits the structure,
— is numerically stable,
— is efficient.

For the computation of deflating subspaces of $\alpha\, A - \beta\, B$, $\alpha\, A' - \beta\, B'$ to find a method that satisfies all these properties is still an open problem. A step in the direction of constructing such a method was recently made by Ammar/Mehrmann [A 4]. It can be based on two reliable numerical methods which exploit the structure and are efficient, of Van Loan [V 7] for the continuous case and Lin [L 21] for the discrete case, which compute the eigenvalues of $\alpha\, A - \beta\, B$, $\alpha\, A' - \beta\, B'$ but not the deflating subspaces. We discuss these two methods in Section 17.

In Section 18 we then discuss the approach of Ammar/Mehrmann [A 4]. The idea of their method is to compute the required deflating subspaces on the basis of the knowledge of the eigenvalues, but with a method that has all the required properties.

All subspace methods can also be used to solve the boundary value problems (1.9), (1.10) directly, using solvers for differential algebraic equations, thus they serve in particular as methods for the cases $K, T < \infty$. Another approach to solve the control problem for the case, $K, T < \infty$ is to solve the Riccati differential equations or difference equations (1.17), (1.20) directly, e.g. Bender/Laub [B 12], [B 13], Laniotis [L 6], Kunkel/Mehrmann [K 26], [K 27]. This topic is discussed in Section 19.

All the presented methods are combined in Section 20 into one big 'expert system' for linear quadratic control problems. Along the lines of a decision flow chart, the reader is led to the algorithm of choice for the particular system properties.

§ 2 Notation and Definitions.

In this section we will introduce the standard notation and the necessary definitions. Some further definitions will be introduced lateron, at the time when they are needed.

We begin with a table of symbols.

2.1 Table: In the following we denote by

$\mathbf{R}$	—	the real numbers,
$\mathbf{R}^{n,m}$	—	the real $n \times m$ matrices, $\mathbf{R}^{n,1} =: \mathbf{R}^n$,
$\mathbf{C}$	—	the complex numbers,
$\mathbf{C}^{n,m}$	—	the complex $n \times m$ matrices, $\mathbf{C}^{n,1} =: \mathbf{C}^n$,
e_j	—	the j-th unit vector,
$I_n(I)$	—	the identity matrix in $\mathbf{C}^{n,n}$, the n is omitted if the dimension is clear.

We use the following matrix (vector) operations:

2.2 Table: Let $A = [a_{ij}] \in \mathbf{K}^{n,n}$ (where $\mathbf{K} = \mathbf{R}$ or $\mathbf{K} = \mathbf{C}$) then we denote by

A^T	—	the transpose of A ,
$\overline{A}$	—	the complex conjugate of A ,
A^*	—	the complex conjugate and transpose of A, $A^* = \overline{A}^T$,
A^{-1}	—	the inverse of A , if A is invertible,
A^+	—	the Moore–Penrose inverse of A , i.e. the matrix $X \in \mathbf{K}^{n,n}$ satisfying $AXA = A$, $XAX = X$, $(AX)^* = AX$, $(XA)^* = XA$,
a_{*j}	—	the j-the column of A ,
a_{j*}	—	the j-th row of A ,
$det(A)$	—	the determinant of A ,
$rk(A)$	—	the rank of A ,
$\sigma(A)$	—	the spectrum of A , i.e. the set of eigenvalues of A ,
$\rho(A)$	—	the spectral radius of A , i.e. $\rho(A) = \max\{|\lambda| \,|\lambda \in \sigma(A)\}$,
$diag(A_1, \ldots, A_k)$	—	the block diagonal matrix with diagonal blocks $A_1, \ldots, A_k$,
$cond_2(A)$	—	the condition of A , $cond_2(A) = \sqrt{\rho(A^*A)} \cdot \sqrt{\rho(A^{-1}A^{-*})}$, $cond_2(A) = \infty$ if A is singular, here $A^{-*} = A^{-1*}$,
$\mathcal{N}(A)$	—	the (right) nullspace of A .

We use the following types of matrices:

2.3 Table: $A = [a_{ij}] \in K^{n,n}$ $(K = R$ or $K = C)$ is said to be

symmetric	— if $a_{ij} = a_{ji}$ $i,j = 1,\dots,n$,
Hermitian	— if $a_{ij} = \overline{a_{ji}}$ $i,j = 1,\dots,n$,
positive semidefinite	— if $x^* A x \geq 0$ for all $x \in K^n$,
positive definite	— if $x^* A x > 0$ for all $x \in K^n \setminus \{0\}$,
nilpotent	— if $A^k = 0$ for some $k \in \{1,\dots,n\}$,
diagonal	— if $a_{ij} = 0$ for $i \neq j$; $A \stackrel{\triangle}{=} [\diagbox]$,
lower triangular	— if $a_{ij} = 0$ for $i < j$; $A \stackrel{\triangle}{=} [\diagbox]$,
upper triangular	— if $a_{ij} = 0$ for $i > j$; $A \stackrel{\triangle}{=} [\diagbox]$,
upper Hessenberg	— if $a_{ij} = 0$ for $i > j+1$; $A \stackrel{\triangle}{=} [\diagbox]$,
quasiupper triangular	— if A is upper Hessenberg and if $a_{i+1,i} \neq 0$ then $a_{i+2,i+1}$, $a_{i,i-1} = 0$,
unreduced upper Hessenberg	— if $a_{ij} = 0$ for $i > j+1$, $a_{ij} \neq 0$ for $i = j+1$,
tridiagonal	— if $a_{ij} = 0$ for $i > j+1$, $i < j - 1$; $A \stackrel{\triangle}{=} [\diagbox]$,
unitary	— if $K = C$ and $A^* A = I$,
orthogonal	— if $A^T A = I$,
c–stable	— if $\sigma(A) \subset \{a \in C \mid \mathcal{R}e(a) < 0\}$ where $\mathcal{R}e(a)$ denotes the real part of a ,
d–stable	— if $\sigma(A) \subset \{a \in C \mid |a| < 1\}$ where $|a|$ denotes the absolute value of a .

2.4 Table: $A = [a_{ij}] = \begin{bmatrix} A_{11} & A_{12} \\ A_{21} & A_{22} \end{bmatrix} \in K^{2n,2n}$, $A_{11}, A_{12}, A_{21}, A_{22} \in K^{n,n}$
$(K = R$ or $K = C)$ is said to be

Hamiltonian	— if $AJ = (AJ)^*$, where $J = \begin{bmatrix} 0 & I_n \\ -I_n & 0 \end{bmatrix}$,
symplectic	— if $AJA^* = J$,
Hamiltonian Hessenberg	— if A Hamiltonian, $A_{11} = -A_{22}^*$ upper Hessenberg and $A_{21} = \alpha\, e_n e_n^*$; $A \stackrel{\triangle}{=} \begin{bmatrix} \diagbox & \square \\ * & \diagbox \end{bmatrix}$,
symplectic Hessenberg	— if A symplectic, A_{11} upper Hessenberg and $A_{21} = p\, e_n^*$ for some $p \in K^n$; $A \stackrel{\triangle}{=} \begin{bmatrix} \diagbox & \square \\ | & \square \end{bmatrix}$,

| Hamiltonian triangular | — | if A Hamiltonian, $A_{11} = -A_{22}^*$ upper triangular and |

$$A_{21} = 0; \quad A \stackrel{\triangle}{=} \begin{bmatrix} \text{} & \text{} \\ \text{} & \text{} \end{bmatrix},$$

| symplectic triangular | — | if A symplectic, $A_{11} = A_{22}^{-*}$ upper triangular |

and $A_{21} = 0$; $A \stackrel{\triangle}{=} \begin{bmatrix} \text{} & \text{} \\ \text{} & \text{} \end{bmatrix},$

| Hamiltonian quasitriangular | — | if A Hamiltonian, $A_{11} = A_{22}^{-*}$ quasi upper triangular, $A_{21} = 0$, |

| symplectic quasitriangular | — | if A symplectic, $A_{11} = A_{22}^{-*}$ quasi upper triangular, $A_{21} = 0$, |

| J–Hessenberg | — | if A_{11}, A_{21}, A_{22} upper triangular and |

A_{21} upper Hessenberg; $A \stackrel{\triangle}{=} \begin{bmatrix} \text{} & \text{} \\ \text{} & \text{} \end{bmatrix},$

| J–triangular | — | if $A_{11}, A_{21}, A_{22}, A_{12}$ upper triangular and |

the diagonal of A_{21} is zero; $A \stackrel{\triangle}{=} \begin{bmatrix} \text{} & \text{} \\ \text{} & \text{} \end{bmatrix},$

| J–tridiagonal | — | if A_{11}, A_{21}, A_{22} diagonal and |

A_{12} tridiagonal; $A \stackrel{\triangle}{=} \begin{bmatrix} \text{} & \text{} \\ \text{} & \text{} \end{bmatrix}.$

A pencil $\alpha A - \beta B$, with $A, B \in K^{2n,2n}$ is called

symplectic — if $AJA^* = BJB^*$,

Hamiltonian — if $AJB^* = BJA^*$.

With this notation we define the following sets of matrices:

2.5 DEFINITION. *We denote by*

$\mathcal{U}_n(\mathbb{C})$	—	*the group of unitary matrices in* $\mathbb{C}^{n,n}$,
$\mathcal{U}_n(\mathbb{R})$	—	*the group of orthogonal matrices in* $\mathbb{R}^{n,n}$,
$\mathcal{S}_{2n}(\mathbb{C})$	—	*the group of symplectic matrices in* $\mathbb{C}^{2n,2n}$,
$\mathcal{S}_{2n}(\mathbb{R})$	—	*the group of symplectic matrices in* $\mathbb{R}^{2n,2n}$,
$\mathcal{US}_{2n}(\mathbb{C})$	—	*the group* $\mathcal{U}_{2n}(\mathbb{C}) \cap \mathcal{S}_{2n}(\mathbb{C})$,
$\mathcal{US}_{2n}(\mathbb{R})$	—	*the group* $\mathcal{U}_{2n}(\mathbb{R}) \cap \mathcal{S}_{2n}(\mathbb{R})$.

We will employ these groups to perform decompositions of matrices. We use the following notation:

2.6 DEFINITION. *Let* $A \in K^{m,m}$ $(K = R$ *or* $K = C)$.

i) A decomposition $A = QR$ *with* $Q \in \mathcal{U}_m(K)$, $R = [r_{ij}] \in K^{m,m}$ *upper triangular,* $r_{ii} \in R$, $r_{ii} > 0$, $i = 1, \dots, m$ *is called QR-decomposition.*

ii) If $m = 2n$, *then a decomposition* $A = SR$ *with* $S \in S_{2n}(K)$, $R \in K^{2n,2n}$ $J-$ *triangular is called SR-decomposition.*

iii) A decomposition $A = U\Sigma V^T$ *with* $U, V \in \mathcal{U}_m(K)$, $\Sigma = \mathrm{diag}(\sigma_1, \dots, \sigma_m)$, $\sigma_1 \geq \dots \geq \sigma_m \geq 0$ *is called singular value decomposition of* A.

We make frequent use of the Jordan normal form of a matrix $A \in C^{n,n}$ and the Kronecker canonical form of a matrix pencil $\alpha A - \beta B \in C^{n,m}$, e.g. Gantmacher [G 4], [G 5].

2.7 THEOREM.

Let $A, B \in C^{n,m}$, *then there exist matrices* $P \in C^{n,n}$, $Q \in C^{m,m}$ *nonsingular such that*

(2.8)
$$P(\alpha A - \beta B)Q =$$
$$\mathrm{diag}\,(\mathcal{O}_s, L_{\epsilon_1}, \dots, L_{\epsilon_p}, L^p_{\nu_1}, \dots, L^p_{\nu_q}, N_{\rho_1}, \dots, N_{\rho_r}, J_{\sigma_1}, \dots, J_{\sigma_s}),$$

where $\mathcal{O}_s$ *is the* $s \times s$ *zero pencil,*

a) L_{ϵ_j} *is an* $\epsilon_j \times (\epsilon_j + 1)$ *bidiagonal pencil of the form*

(2.9)
$$\alpha \begin{bmatrix} 0 & 1 & & 0 \\ \vdots & & \ddots & \\ 0 & 0 & & 1 \end{bmatrix} - \beta \begin{bmatrix} 1 & 0 & & 0 \\ & \ddots & & \vdots \\ & & 1 & 0 \end{bmatrix}$$

b) $L^p_{\nu_j}$ *is an* $(\nu_j + 1) \times \nu_j$ *bidiagonal pencil of the form*

(2.10)
$$\alpha \begin{bmatrix} 1 & & 0 \\ & \ddots & \\ 0 & & 1 \\ 0 & \cdots & 0 \end{bmatrix} - \beta \begin{bmatrix} 0 & \cdots & 0 \\ 1 & & 0 \\ & \ddots & \\ 0 & & 1 \end{bmatrix}$$

c) N_{ρ_j} *is an* $\rho_j \times \rho_j$ *nilpotent pencil*

(2.11)
$$\alpha \begin{bmatrix} 0 & 1 & & \\ & \ddots & \ddots & \\ & & \ddots & 1 \\ & & & 0 \end{bmatrix} - \beta \begin{bmatrix} 1 & & 0 \\ & \ddots & \\ & & \ddots \\ 0 & & & 1 \end{bmatrix}$$

d) J_{σ_j} is a $\sigma_j \times \sigma_j$ Jordan block

(2.12)
$$
\alpha \begin{bmatrix} 1 & & \\ & \ddots & \\ & & \ddots \\ & & & 1 \end{bmatrix} - \beta \begin{bmatrix} \lambda_j & 1 & & \\ & \ddots & \ddots & \\ & & \ddots & 1 \\ & & & \lambda_j \end{bmatrix}
$$

The blockdiagonal pencil (2.8) is called the Kronecker canonical form (KCF) of $\alpha A - \beta B$.

2.13 DEFINITION. $\alpha A - \beta B$ is called regular, if $n = m$ and no Kronecker blocks of type (2.9), (2.10) occur in Kronecker canonical form.

Pairs (α, β) such that $\alpha - \beta \lambda_j = 0$ for λ_j from (2.12) are called (generalized) eigenvalues of $\alpha A - \beta B$ and denoted by λ_j . (Clearly for $t \neq 0$, $(t\alpha, t\beta)$ represents the same eigenvalue, so we define in the following (α, β) to represent $(t\alpha, t\beta)$ for all $t \in \mathbb{C}\backslash\{0\}$.)

Pairs $(\alpha, 0)$, $\alpha \neq 0$ such that $\det (\alpha N_{\rho_j} - \beta I) = 0$ are called infinite eigenvalues of $\alpha A - \beta B$.

The set of generalized and infinite eigenvalue (α, β) of a pencil $\alpha A - \beta B$ is denoted by $\sigma(A, B)$.

By $\text{ind}_\infty (A, B)$ we denote the size of the largest block N_{ρ_j} , the index of $\alpha A - \beta B$.

Two pencils $\alpha A - \beta B$, $\alpha A_1 - \beta B_1 \in \mathbb{C}^{n,m}$ are called equivalent if there exist nonsingular $P \in \mathbb{C}^{n,n}$, $Q \in \mathbb{C}^{m,m}$ such that

$$
P(\alpha A - \beta B)Q = \alpha A_1 - \beta B_1.
$$

If in Definition 2.13 $A = I$ then with $P = Q^{-1}$ the Kronecker canonical form reduces to the usual Jordan canonical form (JCF) of B consisting only of blocks of type (2.12).

2.14 DEFINITION. A k –dimensional subspace χ of $\mathbb{C}^n$ is called

— invariant subspace for $A \in \mathbb{C}^{n,n}$, if there exist matrices $P \in \mathbb{C}^{n,k}$ and $Q \in \mathbb{C}^{k,k}$ such that

(2.15)
$$
AP = PQ
$$

and the columns of P span χ ;

— deflating subspace for $\alpha A - \beta B \in \mathbb{C}^{n,n}$, if there exist matrices $P_1, P_2 \in \mathbb{C}^{n,k}$ and Q_1 , $Q_2 \in \mathbb{C}^{k,k}$ such that

(2.16)
$$
AP_1 = P_2 Q_1, \ BP_1 = P_2 Q_2
$$

and the columns of P_1 span χ .

A subspace $\mathcal{Q}$ of $\mathbb{R}^{2n}$ is called isotropic if $x^T J y = 0$ for all $x, y \in \mathcal{Q}$.

A Lagrangian subspace is a maximal isotropic subspace (i.e., an isotropic subspace that is not contained in a larger isotropic subspace).

Following matrices and matrix pencils, we now discuss triples of matrices (E, A, B) .

In the context of the construction of optimal feedback controls we need the concept of stabilizability and detectability, and their generalizations to descriptor systems. There is a large variety of different definitions in the literature on this topic. Here we use the following notation due to Bunse–Gerstner et al [B 33], [B 34].

2.17 DEFINITION. Let $E, A \in K^{n,n}$, $B \in K^{n,m}$, $C \in K^{p,n}$.

i) The triple (E, A, B) and the corresponding descriptor system is called strongly stabilizable if E, A, B satisfy the following conditions:

$$(2.18) \qquad rk[\lambda E - A, B] = n \ \text{ for all } \begin{cases} \lambda \in C, \ \mathcal{R}e \ \lambda \geq 0 & \text{in the continuous case} \\ \lambda \in C, \ |\lambda| \geq 1 & \text{in the discrete case} \end{cases}$$

and

$$(2.19) \qquad\qquad rk[E, AS_{E,B}, B] = n,$$

where the columns of $S_{E,B}$ span $\mathcal{N}(E) \cap \mathcal{N}(B^*)$ and $S_{E,B}^* S_{E,B} = I$.

ii) The triple (E, A, C) and the corresponding descriptor system is called strongly detectable, if E, A, B satisfy the following conditions

$$(2.20) \qquad rk \begin{bmatrix} \lambda E - A \\ C \end{bmatrix} = n \ \text{ for all } \begin{cases} \lambda \in C, \ \mathcal{R}e \lambda \geq 0 & \text{in the continuous case} \\ \lambda \in C, \ |\lambda| \geq 1 & \text{in the discrete case} \end{cases}$$

and

$$(2.21) \qquad\qquad rk \begin{bmatrix} E \\ L_{E,C}^* A \\ C \end{bmatrix} = n,$$

where the columns of $L_{E,C}$ span $\mathcal{N}(E^*) \cap \mathcal{N}(C)$ and $L_{E,C}^* L_{E,C} = I$.

Obviously if $E = I$ then conditions (2.19), (2.21) are void, condition (2.18) reduces to the usual concept of stabilizability of a pair $[A, B]$ and condition (2.20) reduces to the usual concept of detectability of a pair $[A, C]$, e.g. Knobloch/Kwakernaak [K 18].

§ 3 Existence of solutions.

In this section we review and extend some of the results on existence of solutions to (1.1), (1.2). If the matrix E is singular, then the usual theorems do not apply directly and we have to impose further assumptions on the system matrices and the set of admissable controls. In particular, we would like to preprocess the system in a form that guarantees unique solutions. A very useful result in this context is the following result, which is in this generality due to Bunse–Gerstner et al [B 33] and which we only cite here.

3.1 THEOREM. Let $E, A \in K^{n,n}$, $B \in K^{n,m}$, $C \in K^{p,n}$.

i) The triple (E, A, B) is strongly stabilizable if and only if there exists an $F \in K^{m,n}$ such that

$$(3.2) \qquad \alpha E - \beta(A + BF) \text{ is regular}$$

$$(3.3) \qquad \text{ind}_{\infty}(E, A + BF) \leq 1$$

and for all $(\alpha, \beta) \in C^2$, $\alpha \neq 0$ such that $\det(\alpha E - \beta(A + BF)) = 0$ we have

$$\begin{cases} Re(\frac{\alpha}{\beta}) < 0 \text{ in the continuous case} \\ \left|\frac{\alpha}{\beta}\right| < 1 \text{ in the discrete case.} \end{cases}$$

ii) The triple (E, A, C) is strongly detectable if and only if there exists $G \in K^{n,p}$ such that

$$(3.4) \qquad \alpha E - \beta(A + GC) \text{ is regular}$$

$$(3.5) \qquad \text{ind}_{\infty}(E, A + GC) \leq 1$$

and for all $(\alpha, \beta) \in C^2$, $\beta \neq 0$ such that $\det(\alpha E - \beta(A + GC)) = 0$, we have

$$\begin{cases} Re(\frac{\alpha}{\beta}) < 0 \text{ in the continuous case} \\ \left|\frac{\alpha}{\beta}\right| < 1 \text{ in the discrete case.} \end{cases}$$

iii) The system given by E, A, B, C is strongly stabilizable and strongly detectable if and only if there exists $K \in K^{m,p}$ such that

$$(3.6) \qquad \alpha E - \beta(A + BKC) \text{ is regular}$$

$$(3.7) \qquad \text{ind}_{\infty}(E, A + BKC) \leq 1$$

and for all $(\alpha, \beta) \in C^2$, $\beta \neq 0$ such that $\det(\alpha E - \beta(A + BKC)) = 0$, we have

$$\begin{cases} Re(\frac{\alpha}{\beta}) < 0 \text{ in the continuous case} \\ \left|\frac{\alpha}{\beta}\right| < 1 \text{ in the discrete case.} \end{cases}$$

PROOF: E.g. [B 33]. ∎

Note that strong stabilizability and detectability is invariant under feedbacks as in (3.2), (3.4), (3.6) , e.g. [B 33], [M 9]. If these conditions do not hold, then a general theory for the existence and uniqueness of (1.1), (1.2) is not completely known. There may be nonunique solutions or no solutions at all. The general theory is currently under investigation. The theory most likely involves the mathematical theory of distributions, e.g. Geerts [G 8] in this context, and needs further concepts of controllability and observability, as those for example discussed in Bunse–Gerstner et al [B 33], [B 34] as well as other concepts like the use of derivative feedback combined with proportional feedback. Since this topic is not at all settled, we restrict ourselves here to the case that has recently been studied in a sufficiently complete way by Bunse–Gerstner et al [B 33]. We assume in the following that (E, A, B) is strongly stabilizable and (E, A, C) is strongly detectable.

3.8 COROLLARY. *Let* (E, A, B) *be strongly stabilizable and* (E, A, C) *be strongly detectable. Let*

$$E = U \begin{bmatrix} \Sigma & 0 \\ 0 & 0 \end{bmatrix} V^*$$

be a singular value decomposition of E *with* Σ_E *nonsingular and let* $U^* A V = \begin{bmatrix} A_{11} & A_{12} \\ A_{21} & A_{22} \end{bmatrix}$,
$U^* B = \begin{bmatrix} B_1 \\ B_2 \end{bmatrix}$, $CV = [C_1 \ C_2]$ *be partitioned analogous to* $\begin{bmatrix} \Sigma_E & 0 \\ 0 & 0 \end{bmatrix}$. *Then there exists*
$K \in C^{p,m}$ *such that* $A_{22} + B_2 K C_2$ *is nonsingular.*

PROOF: It is well–known that if (E, A, B) strongly stabilizable and (E, A, B) strongly detectable, then also (UEV^*, UAV^*, UB) is strongly stabilizable and (UE, V^*, UAV^*, CV^*) strongly detectable. By Theorem 3.1 iii) there exists $K \in C^{p,m}$ such that

$$\alpha \tilde{E} - \beta \tilde{A} := \alpha \, UEV^* - \beta(UAV^* + UBKCV^*)$$

(3.9)
$$= \alpha \begin{bmatrix} \Sigma_E & 0 \\ 0 & 0 \end{bmatrix} - \beta \begin{bmatrix} A_{11} + B_1 K C_1 & A_{12} + B_1 K C_2 \\ A_{21} + B_2 K C_1 & A_{22} + B_2 K C_2 \end{bmatrix}$$

$$:= \alpha \begin{bmatrix} \Sigma_E & 0 \\ 0 & 0 \end{bmatrix} - \beta \begin{bmatrix} \tilde{A}_{11} & \tilde{A}_{12} \\ \tilde{A}_{21} & \tilde{A}_{22} \end{bmatrix}$$

is regular and $\text{ind}_\infty(\tilde{E}, \tilde{A}) \leq 1$.

Let now $P, Q \in C^{n,n}$ be nonsingular such that

$$P(\alpha \tilde{E} - \beta \tilde{A})Q = \alpha \begin{bmatrix} I & 0 \\ 0 & 0 \end{bmatrix} - \beta \begin{bmatrix} J_1 & 0 \\ 0 & I \end{bmatrix}$$

is in Kronecker canonical form. If we partition $P^{-1} = \begin{bmatrix} P_{11} & P_{12} \\ P_{21} & P_{22} \end{bmatrix}$, $Q = \begin{bmatrix} Q_{11} & Q_{12} \\ Q_{21} & Q_{22} \end{bmatrix}$
analogous ot $\tilde{E}$, we get from the Kronecker canonical form

$$\alpha \begin{bmatrix} \Sigma_E Q_{11} & \Sigma_E Q_{12} \\ 0 & 0 \end{bmatrix} - \beta \begin{bmatrix} \tilde{A}_{11} & \tilde{A}_{12} \\ \tilde{A}_{21} & \tilde{A}_{22} \end{bmatrix} Q = \alpha \begin{bmatrix} P_{11} & 0 \\ P_{21} & 0 \end{bmatrix} - \beta \begin{bmatrix} P_{11} J_1 & P_{12} \\ P_{21} J_1 & P_{22} \end{bmatrix}$$

thus $Q_{12} = 0$, $P_{21} = 0$, which implies that P_{22}, Q_{22} are nonsingular, since P, Q are nonsingular. But $\tilde{A}_{21}Q_{12} + \tilde{A}_{22}Q_{22} = \tilde{A}_{22}Q_{22} = P_{22}$, hence also $\tilde{A}_{22}$ is nonsingular. ∎

By Theorem 3.1, under the assumptions we have made, we can always transform the system in the following way. Let $K \in \mathbb{C}^{m,p}$ such that

$$\alpha E - \beta(A + BKC) \text{ is regular and } \text{ind } _\infty(E, A + BKC) \leq 1.$$

Consider problem (1.1) and set

(3.10) $$u(t) = Ky(t) + \tilde{u}(t).$$

Then we obtain

$$S(y, u) = \frac{1}{2}\left(y^*(t)My(t)|_{t=T} + \int_{t_0}^{T} [y^*(t)Qy(t) + y^*(t)K^*RKy(t) + y^*(t)K^*R\tilde{u}(t) \right.$$

$$\left. + \tilde{u}^*(t)RKy(t) + \tilde{u}^*(t)R\tilde{u}(t) + y^*(t)SKy(t) + y^*(t)S\tilde{u}(t) + y^*(t)K^*S^*y(t) + \tilde{u}^*(t)S^*y(t)]dt \right)$$

$$= \frac{1}{2}\left(y^*(t)My(t)|_{t=T} + \int_{t_0}^{T} [y^*(t)(Q + K^*RK + SK + K^*S^*)y(t) \right.$$

$$\left. + y^*(t)(S + K^*R)\tilde{u}(t) + \tilde{u}^*(t)(S^* + RK^*)y(t) + \tilde{u}^*(t)R\tilde{u}(t)]dt \right)$$

and
$$E\dot{x}(t) = (A + BKC)x(t) + B\tilde{u}(t)$$
$$y(t) = Cx(t).$$

Setting

(3.11) $$\begin{aligned} A &:= A + BKC, \\ Q &:= Q + K^*RK + SK + K^*S^* \\ S &:= S + K^*R, \end{aligned}$$

we now have a system satisfying

(3.12) $$\alpha E - \beta A \text{ regular and } \text{ind } _\infty(E, A) \leq 1.$$

and still (E, A, B) strongly stabilizable, (E, A, C) strongly detectable.

A numerical procedure to perform this transformation is described in Section 9. For the theoretical part we assume now that the system satisfies (3.12).

In order to discuss the theoretical properties of (1.1) we transform the pencil $\alpha E - \beta A$ into Kronecker canonical form

(3.13) $$\alpha E_1 - \beta A_1 := P_1(\alpha E - \beta A)P_2 = \alpha \begin{bmatrix} I & 0 \\ 0 & 0 \end{bmatrix} - \beta \begin{bmatrix} J_1 & 0 \\ 0 & I \end{bmatrix},$$

with nonsingular $P_1, P_2 \in \mathbb{C}^{n,n}$ and with $J_1 \in \mathbb{C}^{r,r}$. Partitioning

$$(3.14) \quad z(t) := \begin{bmatrix} z_1(t) \\ z_2(t) \end{bmatrix} := P_2^{-1} x(t), \; B_1 := \begin{bmatrix} B_{11} \\ B_{21} \end{bmatrix} := P_1 B, \; C_1 := [C_{11} \quad C_{12}] := C P_2$$

analogous to $P_1 E P_2$ yields the system

$$(3.15) \quad \begin{aligned} \dot{z}_1(t) &= J_1 z_1(t) + B_{11} u(t), & z_1(t_0) &= z_1^0, \\ 0 &= z_2(t) + B_{21} u(t), & z_2(t_0) &= z_2^0, \\ y(t) &= C_{11} z_1(t) + C_{12} z_2(t). \end{aligned}$$

We obtain that $B_{21} u(t_0) = -z_2^0$ as initial condition for u. Thus, the set of admissable controls that we will consider in the following results is

$$U_m^0 = \{u(t) \in \mathbb{C}^m | u(t) \text{ piecewise continuous on } [t_0, T], \; B_{21} u(t_0) = -z_2^0\}.$$

A more general set of functions is also possible, using the general Pontryagin maximum principle, e.g. Pontryagain et al [P 15], but for our purposes U_m^0 is enough. If we go to a distributional setting, less restrictive sets of admissable controls may be considered, e.g. Geerts/Mehrmann [G 11].

3.16 REMARK.

In the following we have to impose another restriction on the system. Let

$$M_1 = P_2^* C^* M C P_2 = \begin{bmatrix} M_{11} & M_{12} \\ M_{21} & M_{22} \end{bmatrix}$$

be partitioned according to (3.14), where P_2 is as in (3.13), then we will require that

$$M_{21}, M_{22}, M_{12} = 0,$$

i.e. the final cost term vanishes on the subspace of the state space given by the algebraic part of (3.13). This assumption is in general not too difficult to meet since in most applications $M = 0$ anyway. We then have the following necessary and sufficient conditions for the existence of the optimal control.

3.17 THEOREM. *Consider control problem (1.1), and assume further that* $\alpha E - \beta A$ *is regular and* $\text{ind}_\infty(E, A) \leq 1$. *Let* $u_* \in U_m^0$ *define the minimal solution of (1.1) and let* $x_* \in \mathbb{C}^n$ *be the corresponding trajectory, i.e. the solution of*

$$(3.18) \quad E\dot{x}(t) = Ax(t) + Bu_*(t), \; x(t_0) = x^0.$$

Let $y_*(t) = Cx_*(t)$ *be the corresponding output function. Assume further, that the final cost term* $y^*(T) My(T) = x^*(T) C^* M C x(T)$ *vanishes on the subspace of the state space defined by the algebraic part of (3.12). (See Remark 3.16).*
Then, there exists a costate function $\mu(t) \in \mathbb{C}^n$, *such that* $x_*(t), \mu(t), u_*(t)$ *satisfy the boundary value problem:*

$$(3.19) \quad \begin{bmatrix} A & 0 & B \\ C^* Q C & A^* & C^* S \\ S^* C & B^* & R \end{bmatrix} \begin{bmatrix} x(t) \\ \mu(t) \\ u(t) \end{bmatrix} = \begin{bmatrix} E & 0 & 0 \\ 0 & -E^* & 0 \\ 0 & 0 & 0 \end{bmatrix} \begin{bmatrix} \dot{x}(t) \\ \dot{\mu}(t) \\ \dot{u}(t) \end{bmatrix}$$

(3.20) $$x(t_0) = x^0, \quad E^*\mu(T) = C^*MCx(T).$$

(Note that $\dot{u}$ is introduced only formally, it does not occur in the equations.)

PROOF: Under the above assumptions, we may assume w.l.o.g. that the Kronecker canonical form of $\alpha E - \beta A$ is as in (3.13) with P_1, P_2 nonsingular and $E_1, A_1, B_1, z(t)$ are as in (3.13), (3.14), (3.15). Let

$$Q_1 := P_2^* C^* Q C P_2, \quad S_1 := P_2^* C^* S, \quad M_1 := P_2^* C^* MC P_2.$$

Replace y by Cx in the cost functional. Then our problem has the form

(3.21) $$\begin{bmatrix} I & 0 \\ 0 & 0 \end{bmatrix} \begin{bmatrix} \dot{z}_1(t) \\ \dot{z}_2(t) \end{bmatrix} = \begin{bmatrix} J_1 & 0 \\ 0 & I \end{bmatrix} \begin{bmatrix} z_1(t) \\ z_2(t) \end{bmatrix} + \begin{bmatrix} B_{11} \\ B_{21} \end{bmatrix} u(t), \quad z(t_0) = z^0$$

and the cost functional is

(3.22) $$S_1(v,u) = \frac{1}{2} \left(z^*(T) M_1 z(T) \right.$$
$$\left. + \int_{t_0}^T (z^*(t) Q_1 z(t) + u^*(t) R u(t) + z^*(t) S_1 u(t) + u^*(t) S_1^* z(t)) dt \right).$$

Let u_*, x_* be the optimal control, trajectory respectively and $z_*(t) = P_2^{-1} x_*(t)$.

Consider a first order perturbation of the optimal control u_*.

(3.23) $$u(t) = u_*(t) + \epsilon v(t).$$

Since $u(t), u_*(t) \in U_m^0$, we have that $v(t)$ has to satisfy $B_{21}v(t_0) = 0$. Substituting (3.23) into (3.21) we obtain for the trajectory corresponding to $u(t)$:

(3.24) $$\dot{z}_1(t) = J_1 z_1(t) + B_{11} v(t) = J_1 z_1(t) + B_{11}(u_*(t) + \epsilon v(t))$$

and using the well–known results giving the solution of (3.24), e.g. Campbell [C 2], we obtain:

$$z_1(t) = e^{J_1(t-t_0)} z^0 + \int_{t_0}^t e^{J_1(t-s)} B_{11}(u_*(s) + \epsilon v(s)) ds$$

$$= z_{*_1}(t) + \epsilon \int_{t_0}^t e^{J_1(t-s)} B_{11} v(s) ds$$

$$z_2(t) = -B_{21} u(t) = -B_{21}(u_*(t) + \epsilon v(t)).$$

Thus, setting

(3.25) $$\varphi(t) = \begin{bmatrix} \int_{t_0}^t e^{J_1(t-s)} B_{11} v(s) ds \\ v(t) \end{bmatrix},$$

we have

(3.26)
$$z(t) = z_*(t) + \epsilon\varphi(t),$$

$\varphi(t_0) = 0$, and $\varphi(t)$ satisfies the differential equation (3.21).

Introducing the costate vector $\mu(t) \in \mathbf{C}^n$ and the Hamilton function $H(z,\mu,u)$ as

(3.27)
$$\begin{aligned}H(z,\mu,u) &= z^*(t)Q_1 z(t) + z^*(t)S_1 u(t) + u^*(t)S_1^* z(t) + u^*(t)Ru(t) \\ &\quad + \mu^*(t)(A_1 z(t) + B_1 u(t)) + (A_1 z(t) + B_1 u(t))^*\mu(t),\end{aligned}$$

then $S_1(z,u)$ takes the form

$$J(u) = \frac{1}{2}\left[\int_{t_0}^{T}(H(z,\mu,u) - \mu^*(t)E_1\dot{z}(t) - \dot{z}^*(t)E_1^*\mu(t))dt \ + \ z(T)^*M_1 z(T)\right]$$

and

$$J(u_*) = \frac{1}{2}\left[\int_{t_0}^{T}(H(z_*,\mu,u_*) - \mu^*(t)E_1\dot{z}_*(t) + \dot{z}_*^*(t)E_1^*\mu(t))dt \ + \ z_*^*(T)M_1 z_*(T)\right].$$

Thus, since $z_*(t_0) = z(t_0)$

(3.28)
$$\begin{aligned}J(u) - J(u_*) = \ &\frac{1}{2}\Bigg[[-z_*^*(t)M_1 z_*(t) + z^*(t)M_1 z(t)]\Big|_{t=T} \\ &+ \int_{t_0}^{T}[H(z,\mu,u) - H(z_*,\mu,u_*)]dt \\ &+ \int_{t_0}^{T}(\mu^*(t)E_1(\dot{z}_*(t) - \dot{z}(t)) + (\dot{z}_*(t) - \dot{z}(t))^*E_1^*\mu(t))dt\Bigg].\end{aligned}$$

We have

$$\begin{aligned}\frac{1}{2}(H(z,\mu,u) - H(z_*,\mu,u_*)) &= \epsilon\{\mathcal{R}e(\varphi^*(t)[Q_1 z_*(t) + S_1 u_*(t) + A_1^*\mu(t)]) \\ &\quad + \mathcal{R}e(\nu^*(t)[B_1^*\mu(t) + S_1^* z_*(t) + Ru_*(t)])\} + \mathcal{O}(\epsilon^2),\end{aligned}$$

and using partial integration,

$$\int_{t_0}^{T}\mathcal{R}e(\mu^*(t)E_1(\dot{z}_*(t) - \dot{z}(t)))dt = -\epsilon\mathcal{R}e(\mu^*(T)E_1\varphi(T)) + \epsilon\int_{t_0}^{T}\mathcal{R}e(\dot{\mu}^*(t)E_1\varphi(t))dt.$$

Thus, (3.28) yields

$$\begin{aligned}J(u) - J(u^*) = \epsilon\Bigg[&\int_{t_0}^{T}[\mathcal{R}e(\varphi^*(t)[Q_1 z_*(t) + S_1 u_*(t) + A_1^*\mu(t) + E_1^*\dot{\mu}(t)]) \\ &+ \mathcal{R}e(\nu^*(t)[B_1^*\mu(t) + S_1^* z_*(t) + Ru_*(t)]) \ dt \\ &+ (\mathcal{R}e(\varphi^*(T)[M_1 z_*(T) - E_1^*\mu(T)]))\Bigg] + \mathcal{O}(\epsilon^2).\end{aligned}$$

Since $\mathcal{J}(u) - \mathcal{J}(u^*) \geq 0$ and ϵ arbitrary, it follows that the factor of ϵ has to vanish. Now we require the costate function μ to satisfy the "initial" value problem

(3.29) $$-F_1^* \dot{\mu}(t) - A_1^* \mu(t) + Q_1 z_*(t) + S_1 u_*(t), \ t \in [t_0, T]$$

(3.30) $$E_1^* \mu(T) = M_1 z_*(T).$$

Observe that due to the special form of E_1, A_1, if we partition $\mu(t)$ accordingly as

$$\begin{bmatrix} \mu_1(t) \\ \mu_2(t) \end{bmatrix}, \ Q_1 \text{ as } \begin{bmatrix} Q_{11} & Q_{12} \\ Q_{21} & Q_{22} \end{bmatrix}, \ S_1 \text{ as } \begin{bmatrix} S_{11} \\ S_{21} \end{bmatrix}, \ M_1 \text{ as } \begin{bmatrix} M_{11} & M_{12} \\ M_{21} & M_{22} \end{bmatrix},$$

we obtain the standard linear system of equations

(3.31) $$-\mu_1(t) = J_1 \mu_1(t) + Q_{11} z_{*_1}(t) + Q_{12} z_{*_2}(t) + S_{11} u_*(t)$$

(3.32) $$\mu_1(T) = M_{11} z_{*_1}(T) + M_{12} z_{*_2}(T)$$

together with the algebraic equation

(3.33) $$0 = \mu_2(t) + Q_{21} z_{*_1}(t) + Q_{22} z_{*_2}(t) + S_{21} u_*(t)$$

(3.34) $$0 = M_{21} z_{*_1}(T) + M_{22} z_{*_2}(T),$$

which is clearly solvable by the standard theory of ordinary differential equations, e.g. [C 2], since we have assumed that $M_{12}, M_{21}, M_{22} = 0$.

So it remains the equality

$$\int_{t_0}^T \mathcal{R}e\left(\nu^*(t)[B_1^* \mu(t) + S_1^* z_*(t) + R\ u_*(t)] \right) dt = 0,$$

which has to hold for all piecewise continuous functions $\nu(t)$ with $\nu(t_0) = 0$.

Now it is simple to show that this implies

(3.35) $$B_1^* \mu(t) + S_1^* z_*(t) + R\ u_*(t) = 0 \text{ on } [t_0, T].$$

Putting (3.35), (3.29), (3.30), (3.21) together and transforming back with P_1, P_2, we obtain the system (3.19), (3.20). ∎

3.36 REMARK.

The proof of Theorem 3.17 is essentially the same proof as in Athans/Falb [A 15] just taking care of the algebraic parts. We see that with a minor restriction, we get the same result. If E is nonsingular then clearly the usual result as in Athans/Falb [A 15] or Sage [S 1] for the maximum principle is obtained by replacing A by $E^{-1}A$ and B by $E^{-1}B$.

The converse of Theorem 3.17 is given in the following Theorem.

3.37 THEOREM. *Suppose x_*, μ, u_* satisfy the boundary value problem (3.19), (3.20) and suppose furthermore that the matrices*

$$\mathcal{R} = \begin{bmatrix} Q & S \\ S^* & R \end{bmatrix}, \ M$$

are positive semidefinite, then

$$S(y, u) = S(Cx, u) \geq S(Cx_*, u_*) = S(y_*, u_*)$$

for all x, y, u satisfying

(3.38)
$$\begin{aligned} E\dot{x}(t) &= Ax(t) + Bu(t), \ x(t_0) = x^0 \\ y(t) &= Cx(t). \end{aligned}$$

PROOF: Following the idea of Campbell [C 2, p. 37], we define $\phi(s) = S(sy_*(t) + (1-s)y(t), su_*(t) + (1-s)u(t))$. Proving Theorem 3.37 is equivalent to showing that $\phi(s)$ has a minimum at $s = 1$ for all $y(t), u(t)$. $\phi(s)$ is quadratic in s, therefore $\phi(s)$ has a minimum at $s = 1$ if and only if

(3.39)
$$\left.\frac{d\phi}{ds}\right|_{s=1} = 0 \text{ and } \left.\frac{d^2\phi}{ds^2}\right|_{s=1} \geq 0.$$

$$\left.\frac{d\phi}{ds}\right|_{s=1} = [(x_*(t) - x(t))^* C^* M C x(t)]|_{t=T}$$

$$+ \int_{t_0}^{T} [(x_*(t) - x(t))^* C^* Q C x_*(t) + x^*(t) C^* Q C (x_*(t) - x(t)) + u^*(t) S^* C (x_*(t) - x(t))$$

$$+ (u_*(t) - u(t))^* S^* C x(t) + (u_*(t) - u(t))^* R u(t) + u^*(t) R (u_*(t) - u(t))$$

$$+ (x_*(t) - x(t))^* C^* S u(t) + x^*(t) C^* S (u_*(t) - u(t))] dt.$$

Multiplying the second equation of (3.19) by $x^*(t)$ and inserting the other equation from (3.19), we obtain

(3.40)
$$\begin{aligned} x_*^*(t) C^* Q C x_*(t) = &-x_*^*(t) E^* \frac{d\mu}{dt} - \left(\frac{d}{dt} x_*(t)\right)^* E^* \mu(t) \\ &+ u_*^*(t)(-S^* C x_*(t) - R u_*(t)) - x_*^*(t) C^* S u_*(t) \end{aligned}$$

and

(3.41)
$$\begin{aligned} x^*(t) C^* Q C x_*(t) = &-x^*(t) E^* \frac{d\mu(t)}{dt} - \left(\frac{d}{dt} x(t)\right)^* E^* \mu(t) \\ &+ u^*(t)(-S^* C x_*(t) - R u_*(t)) - x^*(t) C^* S u_*(t). \end{aligned}$$

Thus, it follows that

$$\frac{d\phi}{ds}\Big|_{s=1} = [(x_*(t) - x(t))^* C^* M C x_*(t)]\,|_{t=T}$$

$$+ \int_{t_0}^{T} (x^*(t) E^* \frac{d\mu(t)}{dt} + \left(\frac{dx(t)}{dt}\right)^* E^* \mu(t) - x_*^*(t) E^* \cdot \frac{d\mu(t)}{dt} - \left(\frac{d}{dt} x_*(t)\right)^* E^* \mu(t)) dt$$

$$= [(x_*(t) - x(t))^* C^* M C x_*(t)]|_{t=T} + [x^*(t) E^* \mu(t)]|_{t_0}^{T} - [x_*^*(t) E^* \mu(t)]|_{t_0}^{T} = 0,$$

since x, x_* satisfy the same initial conditions and $E^* \mu(T) = C^* M C x_*(T)$.

$$\frac{d^2\phi}{ds^2}\Big|_{s=1} = (x_*(t) - x(t))^* C^* M C(x_*(t) - x(t))|_{t=T}$$

$$+ \int_{t_0}^{T} [(x_*(t) - x(t))^*, (u_*(t) - u(t))^*] \begin{bmatrix} C^* Q C & C^* S \\ S^* C & R \end{bmatrix} \begin{bmatrix} x_*(t) - x(t) \\ u_*(t) - u(t) \end{bmatrix} dt$$

but since $M, \begin{bmatrix} Q & S \\ S^* & R \end{bmatrix}$ were assumed positive semidefinite, we have $\frac{d^2\phi}{ds^2}|_{s=1} \geq 0$. ∎

We have given necessary and sufficient conditions for existence of optimal controls for (1.1), which reduce the optimal control problem to the problem of solving a linear boundary value problem. For such problems, the theory is well established, e.g. Campbell [C 2]. If in (3.19) the pencil

$$(3.42) \qquad \alpha \begin{bmatrix} A & 0 & B \\ C^* Q C & A^* & C^* S \\ S^* C & B^* & R \end{bmatrix} - \beta \begin{bmatrix} E & 0 & 0 \\ 0 & -E^* & 0 \\ 0 & 0 & 0 \end{bmatrix} =: \alpha \mathcal{A} - \beta \mathcal{B}$$

is regular, then the general solution is

$$(3.43) \qquad \begin{bmatrix} x(t) \\ \mu(t) \\ u(t) \end{bmatrix} = e^{-\widehat{\mathcal{B}}^D \widehat{\mathcal{A}}(t-t_0)} \widehat{\mathcal{B}}^D \widehat{\mathcal{A}} q,$$

where

$$\widehat{\mathcal{A}} = (\lambda \mathcal{B} - \mathcal{A})^{-1} \mathcal{A}, \quad \widehat{\mathcal{B}} = (\lambda \mathcal{B} - \mathcal{A})^{-1} \mathcal{B}$$

for some λ such that $\det(\lambda \mathcal{B} - \mathcal{A}) \neq 0$. Here $\widehat{\mathcal{B}}^D$ is the Drazin inverse of $\widehat{\mathcal{B}}$, see [C 2].

The solvability of (3.19) then depends only on the consistency of the boundary values (3.20), which may be rewritten as

$$(3.44) \qquad [I \quad 0 \quad 0] \begin{bmatrix} x(t_0) \\ \mu(t_0) \\ u(t_0) \end{bmatrix} = x^0, \quad [-C^* M C \quad E^* \quad 0] \begin{bmatrix} x(T) \\ \mu(T) \\ u(T) \end{bmatrix} = 0.$$

Clearly then the boundary values are consistent if a solution q to the system

$$(3.45) \qquad [I \quad 0 \quad 0]\widehat{\mathcal{B}}^D \widehat{\mathcal{A}} q = x^0, \quad [-C^* M C \quad E^* \quad 0] e^{-\widehat{\mathcal{B}}^D \widehat{\mathcal{A}}(T-t_0)} \widehat{\mathcal{B}}^D \widehat{\mathcal{A}} q = 0$$

exists.

We now discuss the analogous results for the discrete system (1.2). In order to see that for the discrete case the results are analogous, we give the corresponding results in the same order.

3.46 THEOREM. *Consider control problem (1.2) and assume that $\alpha E - \beta A$ is regular and $\mathrm{ind}_{\infty}(E, A) \leq 1$. Let $u_* = (u_{*k_0}, \ldots, u_{*K})$ define the minimal solution of (1.2) and let $x_* = (x_{*k_0}, \ldots, x_{*K})$ define the corresponding states, i.e. the solution of*

$$(3.47) \qquad\qquad Ex_{k+1} = Ax_k + Bu_{*k}, \quad x_{k_0} = x^0.$$

Let $y_{*k} = Cx_{*k}$ be the corresponding outputs. Assume further, that the final cost term $y_K^* M y_K = x_K^* C^* M C x_K$ vanishes on the subspace corresponding to the nilpotent part of $\alpha E - \beta A$ if reduced to Kronecker canonical form (see Remark 3.16). Then, there exist costate vectors $\mu_{k_0}, \ldots, \mu_K \in \mathbb{C}^n$ such that $x_{*k}, \mu_k, u_{*k}, \; k = k_0, \ldots, K$ satisfy the boundary value problem

$$(3.48) \quad \begin{bmatrix} A & 0 & B \\ C^*QC & -E^* & C^*S \\ S^*C & 0 & R \end{bmatrix} \begin{bmatrix} x_k \\ \mu_k \\ u_k \end{bmatrix} = \begin{bmatrix} E & 0 & 0 \\ 0 & -A^* & 0 \\ 0 & -B^* & 0 \end{bmatrix} \begin{bmatrix} x_{k+1} \\ \mu_{k+1} \\ u_{k+1} \end{bmatrix}, \quad k = k_0, \ldots, K-1$$

$$(3.49) \qquad\qquad x_{k_0} = x^0, \; E^*\mu_K = C^* M C x_K.$$

PROOF: The proof is analogous to the proof of Theorem 3.17 if we replace the first order perturbation of the optimal control u_{*k} in (3.23) by

$$(3.50) \qquad\qquad u_k = u_{*k} + \epsilon \nu_k,$$

and introduce the Hamilton function

$$(3.51) \quad \begin{aligned} H(z_k, \mu_k, u_k) &= z_k^* Q_1 z_k + z_k^* S_1 u_k \\ &+ u_k^* S_1^* z_k + u_k^* R u_k + \mu_k^*(A_1 z_k + B_1 u_k) + (A_1 z_k + B_1 u_k)^* \mu_k, \end{aligned}$$

where z_k, μ_k are the discrete analogues to $z(t), \mu(t)$ and Q_1, S_1, A_1, B_1 are as in Theorem 3.17. ∎

3.52 REMARK.

In the discrete case it is possible to prove this result without assuming that $\mathrm{ind}_{\infty}(E, A) \leq 1$. In this case the system depends on future controls, i.e. we have to apply controls $u_k, \; k > K$. See Campbell [C 2] for an analysis of the solutions of system (3.47) if $\mathrm{ind}_{\infty}(E, A) > 1$. From a practical point of view such solutions seem not useful, so we keep the restriction $\mathrm{ind}_{\infty}(E, A) \leq 1$ also in the discrete case.

3.53 THEOREM. *Suppose $x_* = (x_{*k_0}, \ldots, x_{*K}), \; \mu = (\mu_{k_0}, \ldots, \mu_K), \; u_* = (u_{*k_0}, \ldots, u_{*K})$ satisfy the boundary value problem (3.48), (3.49) and suppose further that $\mathcal{R} = \begin{bmatrix} Q & S \\ S^* & R \end{bmatrix}$, M are positive semidefinite, then*

$$S'(y, u) = S'(Cx, u) \geq S'(Cx_*, u_*) = S'(y_*, u_*)$$

for all x, u satisfying

(3.54)
$$E\, x_{k+1} = A\, x_k + B\, u_k, \; k = k_0, \ldots, K-1, \; x_{k_0} = x^0$$
$$y_k = C x_k.$$

PROOF: The proof is completely analogous to the proof of Theorem 3.37 and omitted here. ∎

Analogous to the continuous case, we have reduced the discrete optimal control problem to a linear discrete boundary value problem for which we may apply the well-known theory, e.g. Campbell [C 2]. If in (3.48) the pencil

(3.55)
$$\alpha \begin{bmatrix} A & 0 & B \\ C^*QC & -E^* & C^*S \\ S^*C & 0 & R \end{bmatrix} - \beta \begin{bmatrix} E & 0 & 0 \\ 0 & -A^* & 0 \\ 0 & -B^* & 0 \end{bmatrix} =: \alpha \mathcal{A}' - \beta \mathcal{B}'$$

is regular, then the general solution is

(3.56)
$$\begin{bmatrix} x_k \\ \mu_k \\ u_k \end{bmatrix} = (\widehat{\mathcal{B}}'^D \widehat{\mathcal{A}}')^k \widehat{\mathcal{B}}' \widehat{\mathcal{B}}'^D q$$

where

$$\widehat{\mathcal{A}}' = (\lambda \mathcal{B}' - \mathcal{A}')^{-1} \widehat{\mathcal{A}}', \; \widehat{\mathcal{B}}' = (\lambda \mathcal{B}' - \mathcal{A}')^{-1} \mathcal{B}'$$

for some λ such that $\det(\lambda \mathcal{B}' - \mathcal{A}') \neq 0$, and $\widehat{\mathcal{B}}'^D$ is again the Drazin inverse of $\widehat{\mathcal{B}}'^D$. The solvability of (3.48) then depends only on the consistency of the boundary values (3.49), which may be rewritten as

(3.57)
$$[I \; 0 \; 0] \begin{bmatrix} x_{k_0} \\ \mu_{k_0} \\ u_{k_0} \end{bmatrix} = x^0, \quad [-C^*MC \; E^* \; 0] \begin{bmatrix} x_K \\ \mu_K \\ u_K \end{bmatrix} = 0.$$

Clearly then the boundary values are consistent if a solution to

(3.58)
$$[I \; 0 \; 0] (\widehat{\mathcal{B}}'^D \widehat{\mathcal{A}}')^{k_0} \widehat{\mathcal{B}}' \widehat{\mathcal{B}}'^D q = x^0$$
$$[-C^*MC \; E^* \; 0] (\widehat{\mathcal{B}}'^D \widehat{\mathcal{A}}')^K \widehat{\mathcal{B}}' \widehat{\mathcal{B}}'^D q = 0$$

exists.

We can summarize the results of this section as follows. Under some further assumptions the well-known results on the existence of optimal controls for problems (1.1), (1.2) carry over to the general case, relating the existence of solutions to (1.1), (1.2) to the existence of solutions to the boundary value problems (3.19), (3.20) and (3.48), (3.49) respectively.

These assumptions are guaranteed if (E, A, B) strongly stabilizable and (E, A, C) detectable, since then we can transform the system to a system for which

(3.59)
$$\alpha E - \beta A \; regular$$

(3.60)
$$\text{ind }_\infty(E, A) \leq 1$$

and if furthermore

(3.61) $M_{21}, M_{22}, M_{12} = 0$ for the transformed final cost matrix. See Remark 3.16 .

Under these conditions the optimal solution is given via the solution of a linear two point boundary value problem. For the converse of the existence theorem we need the further conditions:

(3.62)
$$\mathcal{R} = \begin{bmatrix} Q & S \\ S^* & R \end{bmatrix}, \quad M \text{ positive semidefinite.}$$

In the following we assume that conditions (3.59) – (3.62) hold.

§ 4 Eigenstructure of $\alpha A - \beta B$, $\alpha A' - \beta B'$.

In this section we consider the pencils

$$(4.1) \qquad \alpha A - \beta B = \alpha \begin{bmatrix} A & 0 & B \\ C^*QC & A^* & C^*S \\ S^*C & B^* & R \end{bmatrix} - \beta \begin{bmatrix} E & 0 & 0 \\ 0 & -E^* & 0 \\ 0 & 0 & 0 \end{bmatrix}$$

and

$$(4.2) \qquad \alpha A' - \beta B' = \alpha \begin{bmatrix} A & 0 & B \\ C^*QC & -E^* & C^*S \\ S^*C & 0 & R \end{bmatrix} - \beta \begin{bmatrix} E & 0 & 0 \\ 0 & -A^* & 0 \\ 0 & -B^* & 0 \end{bmatrix}$$

corresponding to the homogeneous systems (3.19), (3.48). We discuss conditions under which these are regular pencils, which is a necessary and sufficient condition for the existence of a unique solution of the boundary value problems (3.19), (3.20) and (3.48), (3.49) for all consistent boundary values (e.g. Campbell [C 2]) and we also analyze the eigenstructure of (4.1), (4.2).

If we assume for the moment that R is nonsingular, then $\alpha A - \beta B$, $\alpha A' - \beta B'$ can be transformed equivalently to the pencils

$$(4.3)\ \alpha\mathcal{L} - \beta\mathcal{M} =: \alpha \begin{bmatrix} A - BR^{-1}S^*C & -BR^{-1}B^* & 0 \\ C^*(Q - SR^{-1}S^*)C & A^* - C^*SR^{-1}B^* & 0 \\ S^*C & B^* & R \end{bmatrix} - \beta \begin{bmatrix} E & 0 & 0 \\ 0 & -E^* & 0 \\ 0 & 0 & 0 \end{bmatrix}$$

and
(4.4)

$$\alpha\mathcal{L}' - \beta\mathcal{M}' =: \alpha \begin{bmatrix} A - BR^{-1}S^*C & 0 & 0 \\ C^*(Q - SR^{-1}S^*)C & -E^* & 0 \\ S^*C & 0 & R \end{bmatrix} - \beta \begin{bmatrix} E & BR^{-1}B^* & 0 \\ 0 & -A^* + C^*SR^{-1}B^* & 0 \\ 0 & -B^* & 0 \end{bmatrix}$$

respectively. Clearly these two pencils are regular if and only if the corresponding pencils

$$(4.5) \qquad \begin{aligned} \alpha\tilde{\mathcal{L}} - \beta\widetilde{\mathcal{M}} =: \alpha \begin{bmatrix} F & G \\ H & -F^* \end{bmatrix} - \beta \begin{bmatrix} E & 0 \\ 0 & E^* \end{bmatrix} := \\ \alpha \begin{bmatrix} A - BR^{-1}S^*C & BR^{-1}B^* \\ C^*(Q - SR^{-1}S^*)C & -A^* + C^*SR^{-1}B^* \end{bmatrix} - \beta \begin{bmatrix} E & 0 \\ 0 & E^* \end{bmatrix} \end{aligned}$$

and

$$(4.6) \qquad \begin{aligned} \alpha\tilde{\mathcal{L}}' - \beta\widetilde{\mathcal{M}}' := \alpha \begin{bmatrix} F & 0 \\ H & E^* \end{bmatrix} - \beta \begin{bmatrix} E & -G \\ 0 & F^* \end{bmatrix} \\ := \alpha \begin{bmatrix} A - BR^{-1}S^*C & 0 \\ C^*(Q - SR^{-1}S^*)C & E^* \end{bmatrix} - \beta \begin{bmatrix} E & -BR^{-1}B^* \\ 0 & (A - BR^{-1}S^*C)^* \end{bmatrix}, \end{aligned}$$

respectively, are regular.

These are the forms in which the pencils usually appear in the literature, since in many applications R is positive definite. But there is a great number of applications, where this is not the case, e.g. Emami–Naeini [E 5] for a list of references. For the case $E = I$, pencils (4.5), (4.6) are very well analyzed concerning their eigenstructure since $\alpha \tilde{\mathcal{L}} - \beta \tilde{\mathcal{M}}$ is a Hamiltonian pencil and $\alpha \tilde{\mathcal{L}}' - \beta \tilde{\mathcal{M}}'$ is a symplectic pencil. See for example Laub [L 11], [L 12], Laub/Meyer [14], Lee [L 17], Paige/Van Loan [P 2], Pappas et al [P 5], Singer/Hammarling [S 10] or in a more general setting Gohberg et al [G 14], Lancaster/Rodman [L 4], Mehrmann [M 8]. In the following we will give generalizations of these results, including the cases $E \neq I$, R singular.

4.7 PROPOSITION. Let $\alpha, \beta \in \mathbb{C}$, $(\alpha, \beta) \neq (0,0)$ and let $z^{(j)} \in \mathbb{C}^{2n+m}$, $j = 0, \ldots, k$ such that

$$z^{(j)} = \begin{bmatrix} z_1^{(j)} \\ z_2^{(j)} \\ z_3^{(j)} \end{bmatrix} \quad and$$

(4.8)
$$(\alpha \mathcal{A} - \beta \mathcal{B}) z^{(j)} = \alpha \mathcal{B} z^{(j-1)}, \quad j = 1, \ldots, k,$$

(4.9)
$$(\alpha \mathcal{A} - \beta \mathcal{B}) z^{(0)} = 0,$$

then with $w^{(j)} = \begin{bmatrix} z_2^{(j)} \\ z_1^{(j)} \\ z_3^{(j)} \end{bmatrix}$ we have

(4.10)
$$w^{(j)*}(\overline{\alpha}\mathcal{A} + \overline{\beta}\mathcal{B}) = -\overline{\alpha}\, w^{(j-1)*}\mathcal{B}, \quad j = 1, \ldots, k,$$

(4.11)
$$w^{(0)*}(\overline{\alpha}\mathcal{A} + \overline{\beta}\mathcal{B}) = 0.$$

PROOF:
$$(\alpha \mathcal{A} - \beta \mathcal{B}) \begin{bmatrix} z_1^{(0)} \\ z_2^{(0)} \\ z_3^{(0)} \end{bmatrix} = 0$$

is equivalent to
$$\alpha(A z_1^{(0)} + B z_3^{(0)}) = \beta E^* z_1^{(0)}$$
$$\alpha(C^*QC z_1^{(0)} + A^* z_2^{(0)} + C^*S z_3^{(0)}) = -\beta E^* z_2^{(0)}$$
$$\alpha(S^*C z_1^{(0)} + B^* z_2^{(0)} + R z_3^{(0)}) = 0.$$

Conjugating and transposing these equations, we obtain

(4.12)
$$\overline{\alpha} \begin{bmatrix} z_2^{(0)} \\ z_1^{(0)} \\ z_3^{(0)} \end{bmatrix}^* \begin{bmatrix} A & 0 & B \\ C^*QC & A^* & C^*S \\ S^*C & B^* & R \end{bmatrix} = -\overline{\beta} \begin{bmatrix} z_2^{(0)} \\ z_1^{(0)} \\ z_3^{(0)} \end{bmatrix}^* \begin{bmatrix} E & 0 & 0 \\ 0 & -E^* & 0 \\ 0 & 0 & 0 \end{bmatrix}$$

which is equivalent to (4.11).

Conjugating and transposing (4.8) we get

$$\overline{\alpha}(z_1^{(j)^*} A^* + z_3^{(j)} B^*) - \overline{\beta} z_1^{(j)^*} E^* = \overline{\alpha} z_1^{(j-1)^*} E^*$$
$$\alpha(z_1^{(j)^*} C^* Q C + z_2^{(j)^*} A + z_3^{(j)^*} S^* C) + \overline{\beta} z_2^{(j)^*} E = -\overline{\alpha} z_2^{(j-1)} E$$
$$\alpha(z_1^{(j)^*} C^* S + z_2^{(j)^*} B + z_3^{(j)^*} R) = 0$$

which is equivalent to (4.10). ∎

Observe that Proposition 4.7 includes the cases $\alpha = 0$ or $\beta = 0$, i.e. infinite or zero eigenvalues but not the case of singular pencils.

This is done in the following result:

4.13 PROPOSITION. *Let* $\alpha A - \beta B$ *be a singular pencil. To every chain of vectors* $z^{(0)}, \ldots, z^{(k)}$ *such that*

(4.14) $$Az^{(0)} = 0, \; Az^{(1)} = Bz^{(0)}, \ldots, Az^{(k)} = Bz^{(k-1)}, \; Bz^{(k)} = 0,$$

there exists a corresponding chain $w^{(0)}, \ldots, w^{(k)}$, *where* $w^{(k)} = \begin{bmatrix} z_2^{(k)} \\ z_1^{(k)} \\ z_3^{(k)} \end{bmatrix}$, *with the same partitioning as in Proposition 4.7, is such that*

(4.15) $$w^{(0)^*} A = 0, \ldots, w^{(k)^*} A = w^{(k-1)^*} B, \; w^{(k)^*} = 0.$$

PROOF: Transposing and conjugating (4.14) immediately yields (4.15). ∎

4.16 REMARK.

a) It follows that to every Jordan block

$$\alpha \begin{bmatrix} \lambda & -1 & & \\ & \ddots & \ddots & \\ & & \ddots & -1 \\ & & & \lambda \end{bmatrix} - \beta \begin{bmatrix} 1 & & \\ & \ddots & \\ & & 1 \end{bmatrix}$$

of $\alpha A - \beta B$ for $\lambda \neq -\overline{\lambda}$, there exists a different, paired Jordan block

$$\alpha \begin{bmatrix} -\overline{\lambda} & 1 & & \\ & \ddots & \ddots & \\ & & \ddots & 1 \\ & & & -\overline{\lambda} \end{bmatrix} - \beta \begin{bmatrix} 1 & & \\ & \ddots & \\ & & 1 \end{bmatrix}.$$

b) Observe that if $\alpha A - \beta B$ is a real pencil then we have that to any Jordan block

$$\alpha \begin{bmatrix} \lambda & -1 & & \\ & \ddots & \ddots & \\ & & \ddots & -1 \\ & & & \lambda \end{bmatrix} - \beta \begin{bmatrix} 1 & & \\ & \ddots & \\ & & 1 \end{bmatrix}$$

of $\alpha A - \beta B$ for $\lambda \neq -\lambda$, $\lambda \neq \infty$, there exists a different, paired Jordan block

$$\alpha \begin{bmatrix} -\lambda & 1 & & \\ & \ddots & \ddots & \\ & & \ddots & -1 \\ & & & \lambda \end{bmatrix} - \beta \begin{bmatrix} 1 & & \\ & \ddots & \\ & & 1 \end{bmatrix}.$$

This pairing of blocks does not hold for eigenvalues on the imaginary axis, nor for infinite eigenvalues nor for singular blocks as the following example shows:

4.17 EXAMPLE.

Let

$$\alpha A - \beta B =: \alpha \begin{bmatrix} 0 & 0 & 1 & 0 & 0 & 0 & 0 & 0 & 0 \\ 0 & -i & 0 & 0 & 0 & 0 & 1 & 0 & 0 \\ 0 & 1 & 0 & 0 & 0 & 0 & 0 & 1 & 0 \\ 1 & 0 & 0 & 0 & 0 & 0 & 0 & 0 & 0 \\ 0 & 1 & 0 & 0 & i & 1 & 0 & 1 & 0 \\ 0 & 0 & 0 & 1 & 0 & 0 & 0 & 0 & 0 \\ 0 & 0 & 0 & 0 & 1 & 0 & 1 & 0 & 0 \\ 0 & 1 & 0 & 0 & 0 & 1 & 0 & 1 & 0 \\ 0 & 0 & 0 & 0 & 0 & 0 & 0 & 0 & 0 \end{bmatrix} - \beta \begin{bmatrix} 0 & 0 & 0 & 0 & 0 & 0 & 0 & 0 & 0 \\ 0 & 2 & 0 & 0 & 0 & 0 & 0 & 0 & 0 \\ 0 & 0 & 1 & 0 & 0 & 0 & 0 & 0 & 0 \\ 0 & 0 & 0 & 0 & 0 & 0 & 0 & 0 & 0 \\ 0 & 0 & 0 & 0 & -2 & 0 & 0 & 0 & 0 \\ 0 & 0 & 0 & 0 & 0 & -1 & 0 & 0 & 0 \\ 0 & 0 & 0 & 0 & 0 & 0 & 0 & 0 & 0 \\ 0 & 0 & 0 & 0 & 0 & 0 & 0 & 0 & 0 \\ 0 & 0 & 0 & 0 & 0 & 0 & 0 & 0 & 0 \end{bmatrix}.$$

The Kronecker canonical form of this pencil is as follows:

$$\alpha \begin{bmatrix} 0 & & & & & & \\ 1 & & & & & & \\ & 1 & & & & & \\ & & 1 & & & & \\ & & & 1 & 0 & 0 & \\ & & & 0 & 1 & 0 & \\ & & & 0 & 0 & 1 & \\ & & & & & & i \\ & & & & & & 0 \end{bmatrix} - \beta \begin{bmatrix} 0 & & & & & & \\ & 0 & & & & & \\ & & 0 & & & & \\ & & & 0 & & & \\ & & & & 0 & 0 & 0 \\ & & & & 1 & 0 & 0 \\ & & & & 0 & 1 & 0 \\ & & & & & & 1 \\ & & & & & & & 1 \end{bmatrix}.$$

Clearly there is no pairing for the mentioned types of blocks.

We now give the corresponding results for the discrete case.

4.18 PROPOSITION. Let $(\alpha, \beta) \in \mathbb{C}$, $(\alpha, \beta) \neq (0,0)$ and let $z^{(j)} \in \mathbb{C}^{2n+m}$, $j = 0, \dots, k$ such that

$$z^{(0)} = \begin{bmatrix} z_1^{(j)} \\ z_2^{(j)} \\ z_3^{(j)} \end{bmatrix} \quad \text{and let}$$

(4.19)
$$(\alpha A' - \beta B')z^{(j)} = \alpha B' \, z^{(j-1)},$$

(4.20)
$$(\alpha A' - \beta B')z^{(0)} = 0.$$

Then, for $w^{(j)} = \begin{bmatrix} \alpha\, z_2^{(j)} + \beta\, z_2^{(j-1)} \\ \alpha\, z_1^{(j)} \\ \alpha\, z_3^{(j)} \end{bmatrix}$, $j = 0,\dots,k$, where $z_2^{(-1)} = 0$, we have

(4.21)
$$w^{(j)*}(\bar{\beta}A' - \bar{\alpha}B') = -\bar{\alpha}\, w^{(j-1)*} A',$$

(4.22)
$$w^{(0)*}(\bar{\beta}A' - \bar{\alpha}B') = 0.$$

PROOF: Conjugating and transposing (4.20) we obtain

$$\bar{\alpha}(z_1^{(0)*} A^* + z_3^{(0)*} B^*) = \bar{\beta}\, z_1^{(0)*} E^*$$
$$\bar{\alpha}(z_1^{(0)*} C^* Q C - z_2^{(0)*} E + z_3^{(o)*} S^* C) = -\bar{\beta}\, z_2^{(0)*} A$$
$$\bar{\alpha}(z_1^{(0)*} C^* S + z_3^{(0)*} R) = -\bar{\beta}\, z_2^{(0)*} B.$$

If $\alpha,\beta \neq 0$ then (4.22) follows by multiplying the first equation by $\bar{\alpha}$ and the last two by $\bar{\beta}$.

If $\beta = 0$ then (4.22) is equivalent to $w^{(0)*} B' = 0$, which for the given $w^{(0)} = \begin{bmatrix} 0 \\ \alpha\, z_1^{(j)} \\ \alpha\, z_2^{(j)} \end{bmatrix}$ holds by the first equation of (4.20).

If $\alpha = 0$ then $w^{(0)} = \beta \begin{bmatrix} z_2^{(0)} \\ 0 \\ 0 \end{bmatrix}$ and (4.22) follows by the second equation of (4.20).

To prove (4.21) we first consider $\beta = 0$. Then $w^{(j)} = \alpha \begin{bmatrix} z_2^{(j)} \\ z_1^{(j)} \\ z_3^{(j)} \end{bmatrix}$.

Conjugating and transposing (4.19) we obtain

$$z_1^{(j)*} A^* + z_3^{(j)*} B^* = z_1^{(j-1)*} E^*$$
$$z_1^{(j)*} C^* Q C - z_2^{(j)*} E + z_3^{(j)*} S^* C = -z_2^{(j-1)*} A$$
$$z_1^{(j-1)*} C^* Q C - z_2^{(j-1)*} E + z_3^{(j-1)*} S^* C = -z_2^{(j-2)*} A$$
$$z_1^{(j)*} C^* S + z_3^{(j)*} R = -z_2^{(j-1)*} B$$
$$z_1^{(j-1)*} C^* S + z_3^{(j-1)*} R = -z_2^{(j-2)*} B,$$

which implies that for $j = 1,\dots,k$

$$[z_2^{(j-1)*}, z_1^{(j)*}, z_3^{(j)*}]\, B' = [z_2^{(j-2)*}, z_1^{(j-1)*}, z_3^{(j-1)*}]\, A'$$

and hence (4.21) follows.

If $\alpha = 0$ then $w^{(j)} = \beta \begin{bmatrix} z^{(j)} \\ 0 \\ 0 \end{bmatrix}$ for $j = 0,\dots,k$.

Thus, (4.19) implies $B'z^{(j)} = 0$. But this implies that $w^{(j)^*}A' = 0$ for $j = 0, \ldots, k$.

Now let $\alpha, \beta \neq 0$ and $\lambda = \frac{\beta}{\alpha}$. Then conjugating and transposing (4.19) yields

$$(4.23) \qquad z_1^{(j)^*}A^* + z_3^{(j)^*}B^* - \overline{\lambda}\, z_1^{(j)^*}E^* = z_1^{(j-1)^*}E^*$$

$$(4.24) \qquad z_1^{(j)^*}C^*QC - z_2^{(j)^*}E + z_3^{(j)^*}S^*C + \overline{\lambda}\, z_2^{(j)^*}A = -z_2^{(j-1)^*}A$$

$$(4.25) \qquad z_1^{(j)^*}C^*S + z_3^{(j)^*}R + \overline{\lambda}\, z_2^{(j)^*}B = -z_2^{(j-1)^*}B.$$

Then, to get (4.21), we have to show

$$(4.26) \qquad \begin{aligned} &\overline{\lambda}(\overline{\lambda}\, z_2^{(j)^*} + z_2^{(j-1)^*})A + z_1^{(j)^*}C^*QC + z_3^{(j)^*}S^*C \\ &- (\overline{\lambda}\, z_2^{(j)^*} + z_2^{(j-1)^*})E = -(\overline{\lambda}\, z_2^{(j-1)^*} + z_2^{(j-2)^*})A - z_1^{(j-1)^*}C^*QC - z_3^{(j-1)^*}S^*C \end{aligned}$$

$$(4.27) \qquad -\overline{\lambda}\, z_1^{(j)^*}E^* + z_1^{(j)^*}A^* + z_3^{(j)^*}B^* = z_1^{(j-1)^*}E^*$$

$$(4.28) \qquad \begin{aligned} &\overline{\lambda}[(\overline{\lambda}\, z_2^{(j)^*} + z_2^{(j-1)^*})B + z_1^{(j)^*}C^*S + z_3^{(j)^*}R] \\ &= -[\overline{\lambda}\, z_2^{(j-1)^*} + z_2^{(j-2)^*})B + z_1^{(j-1)^*}C^*S + z_3^{(j-1)^*}R] \end{aligned}$$

(4.26), (4.27) follow by combining two successive equations for indices $j, j-1$ of (4.24), (4.23) respectively and (4.28) follows by combining two successive equations for indices $j, j-1$ of (4.25). ∎

For the singular case we have

4.29 PROPOSITION. *Let $\alpha A' - \beta B'$ be a singular pencil. To every chain of vectors $z^{(0)}, \ldots, z^{(k)} \in \mathbb{C}^{2n+m}$ such that*

$$(4.30) \qquad A'z^{(0)} = 0, \ A'z^{(1)} = B'z^{(0)}, \ldots, A'z^{(k)} = B'z^{(k-1)}, \ B'z^{(k)} = 0$$

there exists a corresponding chain $w^{(0)}, \ldots, w^{(k)}$ such that

$$(4.31) \qquad w^{(0)^*}B' = 0, \ w^{(1)^*}A' = W^{(0)^*}B', \ldots, w^{(k)}B' = w^{(k-1)^*}A', \ w^{(k)^*}A' = 0,$$

where

$$(4.32) \qquad w^{(j)} = \begin{bmatrix} z_2^{(j-1)} \\ z_1^{(j)} \\ z_3^{(j)} \end{bmatrix} \quad \text{and} \quad z_2^{(-1)} = 0.$$

PROOF: $\mathcal{A}'z^{(0)} = 0$ implies $w^{(0)*}\mathcal{B}' = [0, -z_1^{(0)*}A^* - z_3^{(0)*}B^*, 0] = 0$. $\mathcal{A}'z^{(j)} = \mathcal{B}'z^{(j-1)}$ implies that

$$w^{(j-1)*}\mathcal{A}' - w^{(j)*}\mathcal{B}' = [z_2^{(j-2)*}A + z_1^{(j-1)*}C^*QC + z_3^{(j-1)*}S^*C - z_2^{(j-1)*}E,$$
$$- z_1^{(j-1)*}E^* + z_1^{(j)*}A^* + z^{(j)*}B^*, z_2^{(j-2)*}B + z_1^{(j-1)*}C^*S + z_3^{(j-1)*}R] = 0$$

by using successive equations for $j, j - 1$.

$\mathcal{B}'z^{(k)} = 0$ implies that

$$w^{(k)*}\mathcal{A}' = [z_2^{(k-1)*}A + z_1^{(k)*}Q + z_3^{(k)*}S^*C, -z_1^{(k)*}E^*, z_2^{(k-1)*}B + z_1^{(k)*}C^*S + z_3^{(k)*}R] = 0.$$

∎

4.33 REMARK.

a) It follows that to every Kronecker block to an eigenpair $(\alpha, \beta) \neq (\overline{\beta}, \overline{\alpha})$, $\alpha \neq 0$ there exists a Kronecker block of the same size to $(\overline{\beta}, \overline{\alpha})$.

b) If $\alpha\mathcal{A}' - \beta\mathcal{B}'$ is a real pencil, then pairing is for blocks to $(\alpha, \beta) \neq (\beta, \alpha)$, where $\alpha \neq \beta$.

Observe that the pairing for 0 and infinite eigenvalues is not "symmetric" i.e. to every Jordan block to an eigenvalue 0 there is a paired block to an eigenvalue ∞, but not conversely. Also there is not necessarily a pairing for blocks to eigenvalues on the unit circle. An example to show this can be constructed in a similar way as in Example 4.17.

Consequences of the above propositions are the following:

4.34 COROLLARY. Let A, B be as above, and let $z = \begin{bmatrix} z_1 \\ z_2 \\ z_3 \end{bmatrix} \in \mathbb{C}^{n+m}\backslash\{0\}$, $\alpha, \beta \in \mathbb{C}$, $(\alpha, \beta) \neq (0, 0)$ such that $\alpha Az = \beta Bz$. Then,

$$(4.35) \qquad |\alpha|^2 \begin{bmatrix} z_1 \\ z_3 \end{bmatrix}^* \begin{bmatrix} C^*QC & C^*S \\ S^*C & R \end{bmatrix} \begin{bmatrix} z_1 \\ z_3 \end{bmatrix} = -(\alpha\overline{\beta} + \beta\overline{\alpha})z_1^* E^* z_2$$

PROOF: By Proposition 4.7 we have

$$[\alpha\mathcal{A} - \beta\mathcal{B}] \begin{bmatrix} z_1 \\ z_2 \\ z_3 \end{bmatrix} = 0$$

and $[z_2^* \quad z_1^* \quad z_3^*][\overline{\alpha}\mathcal{A} + \overline{\beta}\mathcal{B}] = 0$. This implies

$$|\alpha|^2(z_2^*A^*z_1 + z_1^*C^*QCz_1 + z_3^*S^*Cz_1) = -\alpha\overline{\beta}\, z_2^* Ez_1$$
$$|\alpha|^2(z_2^*A^*z_1 + z_2^*Bz_3) = \overline{\alpha}\beta\, z_2^* Ez_1,$$
$$z_2^* B = -(z_1^*C^*S + z_3^*R).$$

Hence, we have

$$|\alpha|^2(z_1^*Qz_1 + z_3^*S^*Cz_1 + z_1^*C^*Sz_3 + z_3^*Rz_3) = -(\alpha\overline{\beta} + \overline{\alpha}\beta)z_2^* Ez_1$$

which is equivalent to (4.35). ∎

In the discrete case we obtain:

4.36 COROLLARY. Let $\mathcal{A}', \mathcal{B}'$ be as above and let $z = \begin{bmatrix} z_1 \\ z_2 \\ z_3 \end{bmatrix} \in \mathbb{C}^{2n+m} \backslash \{0\}$, $\alpha, \beta \in \mathbb{C}$, $(\alpha, \beta) \neq (0,0)$, such that $\alpha \mathcal{A}'z = \beta \mathcal{B}'z$. Then,

$$
(4.37) \qquad |\alpha|^2 \begin{bmatrix} z_1 \\ z_3 \end{bmatrix}^* \begin{bmatrix} C^*QC & C^*S \\ S^*C & R \end{bmatrix} \begin{bmatrix} z_1 \\ z_3 \end{bmatrix} = -(|\beta|^2 - |\alpha|^2) z_2^* E z_1.
$$

PROOF: If $\beta \neq 0$ then by Proposition 4.18 we have $\begin{bmatrix} \beta z_2 \\ \alpha z_1 \\ \alpha z_3 \end{bmatrix}^* [\bar{\beta} \mathcal{A}' - \bar{\alpha} \mathcal{B}'] = 0$.

This implies

$$
|\beta|^2 \alpha z_2^* (A z_1 + B z_3) = |\beta|^2 \bar{\beta} z_2^* E z_1
$$
$$
\alpha(\bar{\beta}^2 z_1^* A z_1 + \bar{\alpha}\bar{\beta} z_1^* C^*QC z_1 + \bar{\alpha}\bar{\beta} z_3^* S^* C z_1) = \bar{\beta}|\alpha|^2 z_2^* E z_1
$$
$$
\alpha(S^* C z_1 + R z_3) = -\beta B^* z_2.
$$

Hence, we have

$$
|\alpha|^2 \bar{\beta} z_1^* C^*QC z_1 + |\alpha|^2 \bar{\beta} z_3^* S^* C z_1 + \bar{\beta}|\alpha|^2 z_1^* C^* S z_3
$$
$$
+ \bar{\beta}|\alpha|^2 z_3^* R z_3 = -\bar{\beta}(|\beta|^2 - |\alpha|^2) z_2^* E z_1
$$

and this is equivalent to (4.37).

If $\beta = 0$ then $\mathcal{A}'z = 0$. Thus,

$$
A z_1 + B z_3 = 0
$$
$$
C^*QC z_1 - E^* z_2 + C^* S z_3 = 0
$$
$$
S^* C z_1 + R z_3 = 0.
$$

This together implies

$$
\begin{bmatrix} z_1^* & z_3^* \end{bmatrix} \begin{bmatrix} C^*QC & C^*S \\ S^*C & R \end{bmatrix} \begin{bmatrix} z_1 \\ z_3 \end{bmatrix} = z_1^* E z_2 = z_2^* E^* z_1,
$$

since it has to be real. ∎

Corollaries 4.34 and 4.36 generalize known results for Hamiltonian and symplectic matrices. (E.g. Paige/Van Loan [P 2], Mehrmann [M 8])

4.38 COROLLARY. Let A, B be as above and let $z = \begin{bmatrix} z_1 \\ z_2 \\ z_3 \end{bmatrix} \in \mathbb{C}^{2n+m} \backslash \{0\}$, $\alpha, \beta \in \mathbb{C}$, $(\alpha, \beta) \neq (0,0)$, such that $(\alpha A - \beta B)z = 0$.

i) If $\beta = 0$ then $\mathcal{R} = \begin{bmatrix} C^*QC & C^*S \\ S^*C & R \end{bmatrix}$ is singular and $\begin{bmatrix} z_1 \\ z_3 \end{bmatrix} \in \mathcal{N}(\mathcal{R})$.

ii) If $\alpha = 0$ then $z_1 \in \mathcal{N}(E)$, $z_2 \in \mathcal{N}(E^*)$.

iii) If $\alpha \neq 0$, $\mathcal{R}$ positive definite and $\begin{bmatrix} z_1 \\ z_3 \end{bmatrix} \neq 0$, then $\alpha A - \beta B$ has no eigenvalues on the imaginary axis, i.e. $Re\left(\frac{\beta}{\alpha}\right) \neq 0$.

iv) If $\alpha \neq 0$, $Re\left(\frac{\beta}{\alpha}\right) = 0$ then

$$\mathcal{R}\begin{bmatrix} z_1 \\ z_3 \end{bmatrix} = 0, \quad \overline{\alpha} z_2^* A + \overline{\beta} z_2^* E = 0 \text{ and } z_2^* B = 0.$$

v) If (E, A, B) is strongly stabilizable and $\mathcal{R}$ positive definite then $\alpha A - \beta B$ has no eigenvalues on the imaginary axis.

PROOF: i) – iii) follow trivially by Proposition 4.7 and Corollary 4.34.

iv) If $Re\left(\frac{\beta}{\alpha}\right) = 0$ then $\begin{bmatrix} z_1 \\ z_3 \end{bmatrix} \in \mathcal{N}(\mathcal{R})$. Thus, $C^* Q C z_1 + C^* S z_3 = 0$ and $S^* C z_1 + R z_3 = 0$. Hence $B^* z_2 = 0$ and $\overline{\alpha} z_2^* A = -\overline{\beta} z_2^* E$.

v) follows directly by iv) and Definition 2.17. ■

The corresponding discrete result is as follows:

4.39 COROLLARY. Let A', B' be as above and let $z = \begin{bmatrix} z_1 \\ z_2 \\ z_3 \end{bmatrix} \in \mathbf{C}^{2n+m} \backslash \{0\}$, $\alpha, \beta \in \mathbf{C}$, $(\alpha, \beta) \neq (0, 0)$, such that $(\alpha A' - \beta B') z = 0$.

i) If $\beta = 0$ then $\mathcal{R}\begin{bmatrix} z_1 \\ z_3 \end{bmatrix} = \begin{bmatrix} E^* z_2 \\ 0 \end{bmatrix}$.

ii) If $\alpha = 0$ then $E z_1 = 0$.

iii) If $\mathcal{R}$ is positive definite, $\alpha \neq 0$ and $\begin{bmatrix} z_1 \\ z_3 \end{bmatrix} \neq 0$, then $\alpha A' - \beta B'$ has no eigenvalues on the unit circle, i.e. $\left|\frac{\beta}{\alpha}\right| \neq 1$.

iv) If $\alpha \neq 0$, $|\alpha|^2 = |\beta|^2$ then $\mathcal{R}\begin{bmatrix} z_1 \\ z_3 \end{bmatrix} = 0$, $z_2^* B = 0$, $\overline{\beta} z_2^* A = \overline{\alpha} z_2^* E$.

v) If (E, A, B) is strongly stabilizable and $\mathcal{R}$ positive definite, then $\alpha A' - \beta B'$ has no eigenvalues on the unit circle.

PROOF: Analogous to proof of Corollary 4.38. ■

We now discuss further conditions for regularity of the pencils $\alpha A - \beta B$, $\alpha A' - \beta B'$. We begin with obvious necessary conditions.

4.40 PROPOSITION. *If* $\alpha\mathcal{A} - \beta\mathcal{B}$ $(\alpha\mathcal{A}' - \beta\mathcal{B}')$ *is a regular pencil, then* rk $\begin{bmatrix} B \\ C^*S \\ R \end{bmatrix} = m$.

PROOF: If rk $\begin{bmatrix} B \\ C^*S \\ R \end{bmatrix} < m$ then there exists a nonsingular $m \times m$ matrix Q such that

$(\alpha\mathcal{A} - \beta\mathcal{B}) \begin{bmatrix} I & 0 & 0 \\ 0 & I & 0 \\ 0 & 0 & Q \end{bmatrix}$ has last column 0, which contradicts the regularity. ∎

The rank condition of Proposition 4.40 is clearly satisfied if $\mathcal{R}$ is positive definite but this is not necessarily the case in all practical problems, see for example Emami–Naeini [E 5], Willems et at [W 12], Geerts [G 7], [G 8], [G 9], [G 10].

For the solution of the optimal control problems a rank defect in $\begin{bmatrix} B \\ C^*S \\ R \end{bmatrix}$ can be treated

as follows:

Observe that if Q_1 is unitary such that $\begin{bmatrix} B \\ C^*S \\ R \end{bmatrix} Q_1 = \begin{bmatrix} B_1 & 0 \\ C^*S_1 & 0 \\ R_1 & 0 \end{bmatrix}$, where $\begin{bmatrix} B_1 \\ C^*S_1 \\ R_1 \end{bmatrix}$

has full column rank ρ (take for example Q_1 to be the right factor in the singular value

decomposition of $\begin{bmatrix} B \\ C^*S \\ R \end{bmatrix}$, (e.g. Golub/Van Loan [G 12])), then the original control

problem can be reduced to the problem of minimizing

$$(4.41) \qquad S(x,u_1) = \frac{1}{2}(x^*(t)C^*MCx(t)) \Bigg|_{t_0}^T + \int_{t_0}^T (x^*(t)C^*QCx(t)$$
$$+ x^*(t)C^*S_1u_1(t) + u_1^*(t)S_1^*Cx(t) + u_1^*(t)R_1u_1(t))dt)$$

subject to

$$E\dot{x}(t) = Ax(t) + B_1u_1(t), \ x(t_0) = x^0, \ y(t) = Cx(t)$$

where $u_1(t)$ consists of the first ρ components of $Q_1^*u(t)$. Clearly the remaining $m - \rho$ components of $Q_1^*u(t)$ do not influence the system any more and thus can be simply omitted. Thus, for the further analysis of the system we assume in the following that

$$(4.42) \qquad \text{rk} \begin{bmatrix} B \\ C^*S \\ R \end{bmatrix} = m.$$

The same arguments hold for the discrete case.

If rk $\begin{bmatrix} C^*S \\ R \end{bmatrix} < m$ then we have controls, which are costfree, since

$$\int_{t_0}^T (u^*(t)Ru(t) + u^*(t)S^*Cx(t) + x^*(t)C^*Su(t))dt$$

may be zero. There exists an important class of control problems, which can be reformulated to fit into this setting, the so-called deadbeat control problems, e.g. Emami–Naeini [E 5], Emami–Naeini/Franklin [E 6], [E 7]. In these problems one is looking for controls $u(t), u_k$ such that the solution of

$$E\dot{x}(t) = Ax(t) + Bu(t), \quad t \in (t_0, T), \ x(t_0) = x^0$$

or

$$Ex_{k+1} = Ax_k + Bu_k, \quad k = k_0, \dots, K-1, \ x_{k_0} = x^0$$

approaches 0 as fast as possible for increasing t or k. For the fomulation of these problems as problems of type (1.1), (1.2) see Emami–Naeini [E 5] or Emami–Naeini/Franklin [E 7]. A complete discussion of the theory in the case that $\mathcal{R}$ is not positive definite is given by Geerts [G 10].

We do not discuss this topic here further although it leads to algorithms similar to those that we discuss. A sufficient condition for regularity of $\alpha A - \beta B$, $\alpha A' - \beta B'$ is obtained via the following Theorem.

4.43 THEOREM. *Let* (E, A, B) *be strong stabilizable and* $\mathcal{R}$ *positive definite then* $\alpha A - \beta B$ $(\alpha A' - \beta B')$ *is a regular pencil.*

PROOF: Suppose $\alpha A - \beta B$ is singular, then for all $\alpha, \beta \in \mathbb{C}\backslash\{0\}$ with $\mathcal{R}e\left(\frac{\beta}{\alpha}\right) = 0$,

$\left(\left|\frac{\beta}{\alpha}\right| = 1\right)$ there exists $z = \begin{bmatrix} z_1 \\ z_2 \\ z_3 \end{bmatrix} \in \mathbb{C}^{2n+m}\backslash\{0\}$, such that $\alpha Az = \beta Bz$ $(\alpha A'z = \beta B'z)$.

By Corollary 4.34, (Corollary 4.36) it follows that $\begin{bmatrix} z_1 \\ z_3 \end{bmatrix} = 0$. Thus $z_2 \neq 0$, $\alpha z_2^* A = \beta z_2^* E$, $z_2^* B = 0$ for all $\alpha, \beta \in \mathbb{C}\backslash\{0\}$ with $\mathcal{R}e\left(\frac{\beta}{\alpha}\right) = 0$ $\left(\left|\frac{\beta}{\alpha}\right| = 1\right)$.

Then, it follows that

$$\frac{\alpha}{\beta} z_2^*(A + BF) = z_2^* E$$

for an infinite number of eigenvalues $\frac{\alpha}{\beta}$, and for all $F \in \mathbb{C}^{m,n}$. But this contradicts the strong stabilizability of (E, A, B). ∎

We can summarize the results of this section as follows:

The pencils $\alpha A - \beta B$ *and* $\alpha A' - \beta B'$ *inherit most of the spectral properties of the corresponding Hamiltonian and symplectic matrices, respectively. There is a pairing of Jordan blocks for finite eigenvalues, which are not on the imaginary axis or on the unit circle respectively.*

Under the assumptions that $\mathcal{R} = \begin{bmatrix} C^*QC & C^*S \\ S^*C & R \end{bmatrix}$ *positive definite and* (E, A, B) *strongly stabilizable the pencils* $\alpha A - \beta B$ *and* $\alpha A' - \beta B'$ *have no eigenvalues on the imaginary axis and the unit circle, respectively. If* rk $\begin{bmatrix} C^*S \\ R \end{bmatrix} < m$, *then costfree controls exist, and one can impose further conditions to deal with this case.*

If rk $\begin{bmatrix} B \\ C^*S \\ R \end{bmatrix}$ < m then there are arbitrary components in $u(t)$, which do not influence
the problem at all.

§ 5 Uniqueness and stability of feedback solutions.

In Section 3 we have given existence theorems for optimal solutions of (1.1), (1.2). In this section we now discuss uniqueness and stability of solutions.

Observe that if R is nonsingular, then we have

$$(5.1) \qquad u(t) = -R^{-1}(B^*\mu(t) + S^*Cx(t))$$

$$(5.2) \qquad u_k = -R^{-1}(S^*Cx_k - B^*\mu_{k+1}).$$

If E is nonsingular then it is well–known that the optimal controls are feedback controls

$$(5.3) \qquad u(t) = K(t)y(t)$$

$$(5.4) \qquad u_k = K_k y_k$$

for some matrix valued function $K(t)$, $K_k \in \mathbb{C}^{m,p}$.

In this section we generalize these results to the descriptor case and also discuss the stability of the optimally controlled or closed loop system.

We begin with the continuous case and the well–known results in the case that E, R are nonsingular.

To study this question consider first the Riccati differential equation

$$(5.5) \qquad \begin{aligned} -E^*\dot{X}(t)E &= C^*QC + E^*X(t)A + A^*X(t)E \\ &\quad - (B^*X(t)E + S^*C)^*R^{-1}(B^*X(t)E + S^*C) \end{aligned}$$

with the "terminal" condition

$$(5.6) \qquad E^*X(T)E = C^*MC$$

and the algebraic Riccati equation

$$(5.7) \qquad 0 = C^*QC + E^*XA + A^*XE - (B^*XE + S^*C)^*R^{-1}(B^*XE + S^*C).$$

The solvability of (5.5), (5.6) in the case that E is nonsingular is very well studied in Reid [R 4]. The solvability of (5.7) for the case that E is nonsingular is also well–studied, Brocket [B 18], Kalman [K 4], Kučera [K 23], [K 24], [K 25], a recent review is given in Lancaster/Rodman [L 4].

We have the following well–known results.

5.8 THEOREM. Let E be nonsingular, R positive definite, $T < \infty$. Then the optimal solution to problem (1.1) exists, is unique and given by the feedback law:

$$(5.9) \qquad u(t) = -R^{-1}(S^*C + B^*X(t)E)\, x(t),$$

where the Hermitian matrix $X(t)$ is the unique solution of the Riccati differential equation (5.5) with the terminal condition (5.6).

PROOF: If we replace A by $E^{-1}A$, B by $E^{-1}B$, $u(t)$ by $R^{-1}S^*Cx(t)+u(t)$ then the proof is the standard proof as in Athans/Falb [A 15, p. 763]. ∎

In the infinite time case this question is a little more difficult. In general one has to distinguish between optimal, stabilizing and asymptotic solutions for $T \to \infty$.

A summary of the existence and uniqueness results due to Kučera [K 24] in this case is given in Casti [C 4, p. 204], where two nice tables are given. Here we only mention the most important case, e.g. Casti [C 4].

5.10 THEOREM. *Let E be nonsingular, R positive definite, (E, A, B) strongly stabilizable, $(E, A, C^*(Q-SR^{-1}S^*)C)$ strongly detectable. Then the optimal control for problem (1.2) with $T = \infty$ exists, is unique and given by the feedback law.*

(5.11) $$u(t) = -R^{-1}(B^*XE + S^*C)\, x(t),$$

where X is the unique positive definite solution of the algebraic Riccati equation (5.7).

Furthermore the (closed loop) dynamics of the system obtained with this control,

$$E\dot{x}(t) = [A - BR^{-1}(B^*XE + S^*C)]x(t)$$

is asymptotically stable, i.e. $\lim_{t\to\infty} x(t) = 0$.

Note that for E nonsingular the condition that (E, A, B) is strongly stabilizable and (E, A, C) strongly detectable is the usual condition of stabilizability and detectability as for example in Casti [C 4]. In the case that E is singular the situation becomes more complicated. We have to impose further restrictions to obtain unique, stabilizing feedback controls. Major steps in this direction are papers by Cobb [C 9], [C 10] where the case $R = I$, $C^*QC = I$, $S = 0$, $T = \infty$ is analyzed for E singular. Cobb proves (in our notation)

5.12 THEOREM. *If (E, A, B) is strongly stabilizable, $R = I$, $C^*QC = I$, $S = 0$, then problem (1.1) with $T = \infty$ has a unique solution.*

Note that in this case (E, A, C^*QC) is strongly detectable, since $C^*QC = I$. In general one has to require (E, A, C^*QC) strongly detectable to get uniqueness. We will see in the following how these results and others can be obtained by a reduction similar to that used in Bender/Laub [B 12] or in Section 3.

We will begin with some obvious remarks on this point. We have seen in Section 4 that controls, which satisfy

(5.13) $$\begin{bmatrix} B \\ C^*S \\ R \end{bmatrix} u(t) = 0 \quad \text{for all } t \in [t_0, T]$$

or

$$(5.14) \qquad \begin{bmatrix} B \\ C^*S \\ R \end{bmatrix} u_k = 0 \quad \text{for all } k = k_0, \dots, K$$

do not influence the dynamics and are cost free. So in order to have a unique solution we need necessarily

$$(5.15) \qquad \text{rk} \begin{bmatrix} B \\ C^*S \\ R \end{bmatrix} = m.$$

In the following we therefore assume that (5.15) holds and in view of Section 4 furthermore that R is positive definite. It is obvious that a solution to the optimal control problem (1.1) if it exists, is unique if and only if the corresponding solution of the boundary value problem (3.19), (3.20) is unique. To see, when this is the case, we assume that the boundary conditions are consistent and we use the well–known results on the uniqueness of solutions to linear homogenous boundary value problems, e.g. Campbell [C 2].

Assume without loss of generality that that $\alpha E - \beta A$ is in Kronecker canonical form (3.13). Then system (3.19), (3.20) in the notation used in (3.14), (3.31), (3.32), (3.33), (3.34) can be permuted as

$$(5.16) \qquad \begin{bmatrix} I & 0 & 0 & 0 & 0 \\ 0 & -I & 0 & 0 & 0 \\ 0 & 0 & 0 & 0 & 0 \\ 0 & 0 & 0 & 0 & 0 \\ 0 & 0 & 0 & 0 & 0 \end{bmatrix} \begin{bmatrix} \dot{z}_1(t) \\ \dot{\mu}_1(t) \\ \dot{\mu}_2(t) \\ \dot{z}_2(t) \\ \dot{u}(t) \end{bmatrix} = \begin{bmatrix} J_1 & 0 & 0 & 0 & B_{11} \\ Q_{11} & J_1^* & 0 & Q_{12} & S_{11} \\ 0 & 0 & 0 & I & B_{21} \\ Q_{21} & 0 & I & Q_{22} & S_{21} \\ S_{11}^* & B_{11}^* & B_{21}^* & S_{21}^* & R \end{bmatrix} \begin{bmatrix} z_1(t) \\ \mu_1(t) \\ \mu_2(t) \\ z_2(t) \\ u(t) \end{bmatrix}$$

$$(5.17) \qquad z_1(t_0) = z_1^0, \ z_2(t_0) = z_2^0, \ \mu_1(T) = M_{11} z_1(T).$$

To discuss the question of uniqueness, we first consider the system

$$(5.18) \qquad \begin{bmatrix} 0 & I & B_{21} \\ I & Q_{22} & S_{21} \\ B_{21}^* & S_{21}^* & R \end{bmatrix} \begin{bmatrix} \mu_2(t) \\ z_2(t) \\ u(t) \end{bmatrix} = - \begin{bmatrix} 0 & 0 \\ Q_{21} & 0 \\ S_{11}^* & B_{11}^* \end{bmatrix} \begin{bmatrix} z_1(t) \\ \mu_1(t) \end{bmatrix}.$$

Performing a block elimination to block triangular form, we obtain

$$(5.19) \qquad \begin{bmatrix} I & 0 & B_{21} \\ 0 & I & S_{21} - Q_{22}B_{21} \\ 0 & 0 & \tilde{R} \end{bmatrix} \begin{bmatrix} z_2(t) \\ \mu_2(t) \\ u(t) \end{bmatrix} = - \begin{bmatrix} 0 & 0 \\ Q_{21} & 0 \\ S_{11}^* - B_{21}^* Q_{21} & B_{11}^* \end{bmatrix} \begin{bmatrix} z_1(t) \\ \mu_1(t) \end{bmatrix},$$

where $\tilde{R} = R - S_{21}^* B_{21} - B_{21}^* S_{21} + B_{21}^* Q_{22} B_{21} = [-B_{21}^* \quad I] \begin{bmatrix} Q_{22} & S_{21} \\ S_{21}^* & R \end{bmatrix} \begin{bmatrix} -B_{21} \\ I \end{bmatrix}$ is

positive definite, since $\begin{bmatrix} Q_{22} & S_{21} \\ S_{21}^* & R \end{bmatrix}$ is a principle submatrix of a positive definite matrix,

and $\begin{bmatrix} -B_{21} \\ I \end{bmatrix}$ has full column rank.

Solving the last equation in (5.19) for $u(t)$ and inserting in (5.16), we obtain the system

(5.20)
$$\dot{z}_1(t) = [J_1 - B_{11}\tilde{R}^{-1}(S_{11}^* - B_{21}^*Q_{21})]z_1(t) - B_{11}\tilde{R}^{-1}B_{11}^*\mu_1(t)$$
$$\dot{\mu}_1(t) = [-Q_{11} + (S_{11} - Q_{12}B_{21})\tilde{R}^{-1}(S_{11}^* - B_{21}^*Q_{21})]z_1(t)$$
$$+ [-J_1^* + (S_{11} - Q_{12}B_{21})\tilde{R}^{-1}B_{11}^*)]\mu_1(t)$$

or in matrix form

(5.21)
$$\begin{bmatrix} \dot{z}_1(t) \\ \dot{\mu}_1(t) \end{bmatrix} = \mathcal{H} \begin{bmatrix} z_1(t) \\ \mu_1(t) \end{bmatrix},$$

where
(5.22)
$$\mathcal{H} = \begin{bmatrix} J_1 - B_{11}\tilde{R}^{-1}(S_{11}^* - B_{21}^*Q_{21}) & -B_{11}\tilde{R}^{-1}B_{11}^* \\ -Q_{11} + (S_{11} - Q_{12}B_{21})\tilde{R}^{-1}(S_{11}^* - B_{21}^*Q_{21}) & -J_1^* + (S_{11} - Q_{12}B_{21})\tilde{R}^{-1}B_{11}^* \end{bmatrix}$$

is Hamiltonian in $\mathbb{C}^{2r,2r}$. The general solution of (5.21) is

(5.23)
$$\begin{bmatrix} z_1(t) \\ \mu_1(t) \end{bmatrix} = e^{\mathcal{H}(t-t_0)} \begin{bmatrix} q_1 \\ q_2 \end{bmatrix}.$$

The solution with boundary conditions (5.17), if it exists, is unique if and only if the homogeneous system (5.21) with boundary conditions

(5.24)
$$z_1(t_0) = z_1^0, \ \mu_1(T) - M_{11}z_1(T) = 0$$

is uniquely solvable.

Let $\tilde{\mathcal{H}} = e^{\mathcal{H}(T-t_0)} = \begin{bmatrix} \tilde{H}_{11} & \tilde{H}_{12} \\ \tilde{H}_{21} & \tilde{H}_{22} \end{bmatrix}$. It follows that the solution of (5.21) with boundary conditions (5.24) if it exists, is unique if and only if the system

(5.25)
$$(M_{11}\tilde{H}_{12} - \tilde{H}_{22})q_2 = (\tilde{H}_{21} - M_{11}\tilde{H}_{11})z_1^0$$

is uniquely solvable, i.e. $M_{11}\tilde{H}_{12} - \tilde{H}_{22}$ nonsingular. We can summarize these results as follows:

5.26 THEOREM. If $\alpha E - \beta A$ is regular, $\text{ind}_\infty(E, A) \leq 1$ and the boundary conditions (3.20) are consistent, then the solution of the boundary value problem (3.19), (3.20) is unique if and only if $M_{11}\tilde{H}_{12} - \tilde{H}_{22}$ is invertible.

Note that this result is essentially stated in Bender/Laub [B 12], where also system theoretic interpretations of these invertibility conditions are given. Note further that for $M = 0$, Theorem 5.26 implies that the solution is unique if and only if $\tilde{H}_{22}$ is nonsingular.

Since R is assumed nonsingular, (5.16) can be rewritten as the system

(5.27)
$$\begin{bmatrix} I & 0 & 0 \\ 0 & -I & 0 \\ 0 & 0 & 0 \end{bmatrix} \begin{bmatrix} \dot{z}_1(t) \\ \dot{\mu}_1(t) \\ \dot{u}(t) \end{bmatrix} = \begin{bmatrix} J_1 & 0 & B_{11} \\ Q_{11} & J_1 & S_{11} - Q_{12}B_{21} \\ S_{11}^* - B_{21}^*Q_{21} & B_{11}^* & R \end{bmatrix} \begin{bmatrix} z_1(t) \\ \mu_1(t) \\ u(t) \end{bmatrix}$$

with boundary conditions

(5.28) $$z_1(t_0) = z_1^0, \ \mu_1(T) = M_{11}z_1(T), \ -B_{21}u(t_0) = z_2^0.$$

This system may be viewed as the linear boundary value problem stemming from the control problem

(5.29) $$\begin{aligned}\text{Minimize } \ \widetilde{S}(z_1, u) = &\frac{1}{2}[(z_1(T)^* M_{11} z_1(T) - z_1^*(t_0)M_{11}z_1(t_0) \\ &+ \int_{t_0}^{T} [z_1^*(t)Q_{11}z_1(t) + z_1^*(t)(S_{11} - Q_{12}B_{21})u(t) \\ &+ u^*(t)(S_{11}^* - B_{21}^* Q_{21})z_1(t) + u^*(t)\widetilde{R}u(t)] \ dt)]\end{aligned}$$

subject to the constraints

(5.30) $$\dot{z}_1(t) = J_1 z_1(t) + B_{11}u(t), \ z_1(t_0) = z_1^0, \ -B_{21}u(t_0) = z_2^0.$$

Note that if (E, A, B) is strongly stabilizable, then also (I, J_1, B_1) is strongly stabilizable, since the transformation to Kronecker canonical form does not change the stabilizability and also $(I, J_1, Q_{11} - S_{11}\widetilde{R}^{-1}S_{11}^*)$ is strongly detectable if $(E, A, C^*(Q - SR^{-1}S^*)C)$ was strongly detectable. See Mehrmann/Krause [M 9].

Thus, we have (I, J_1, B_1) strongly stabilizable and $(I, J_1, Q_{11} - S_{11}\widetilde{R}^{-1}S_{11}^*)$ strongly detectable. Applying Theorem 5.8, the optimal feedback for (5.29), (5.30) is given by

(5.31) $$u(t) = -\widetilde{R}^{-1}(B_{11}^* X + (S_{11} - Q_{12}B_{21}))z_1(t)$$

where X is the positive semidefinite solution of the algebraic Riccati equation

(5.32) $$0 = Q_{11} + XJ_1 + J_1^* X - (B_{11}^* X + S_{11} - Q_{12}B_{21})^* \widetilde{R}^{-1}(B_{11}^* X + S_{11} - Q_{12}B_{21}).$$

The condition $-B_{21}u(t_0) = z_2^0$, which has to be fulfilled, is then given by

(5.33) $$B_{21}\widetilde{R}^{-1}(B_{11}^* X + (S_{11} - Q_{12}B_{21}))z_1^0 = z_2^0.$$

The same type of consistency condition is obtained for $T < \infty$ with X replaced by $X(t)$. If we discuss the problem in the distributional setting, often a wider class of initial conditions is possible, e.g. Geerts/Mehrmann [G 11]

Observe further, that under the assumptions of strong stabilizability and strong detectability the complete vector $x(t) = \begin{bmatrix} x_1(t) \\ x_2(t) \end{bmatrix} = P_2 \begin{bmatrix} z_1(t) \\ z_2(t) \end{bmatrix}$ will be asymptotically stable, i.e. satisfy $\lim_{t \to \infty} x(t) = 0$, since $z_2(t) = -B_{21}u(t)$ is just a constant matrix multiplied to $z_1(t)$.

If we analyze the Riccati differential equations (5.5), (5.6) and the algebraic Riccati equation (5.7) for the case that E is singular, we can perform the transformations that transform

$\alpha E - \beta A$ in Kronecker canonical form also with this equation. We then obtain with the notation as above

(5.34)
$$-\begin{bmatrix} I & 0 \\ 0 & 0 \end{bmatrix}\begin{bmatrix} \dot{Y}_{11}(t) & \dot{Y}_{12}(t) \\ \dot{Y}_{21}(t) & \dot{Y}_{22}(t) \end{bmatrix}\begin{bmatrix} I & 0 \\ 0 & 0 \end{bmatrix} = \begin{bmatrix} Q_{11} & Q_{12} \\ Q_{21} & Q_{22} \end{bmatrix}$$
$$+ \begin{bmatrix} I & 0 \\ 0 & 0 \end{bmatrix}\begin{bmatrix} Y_{11}(t) & Y_{12}(t) \\ Y_{21}(t) & Y_{22}(t) \end{bmatrix}\begin{bmatrix} J_1 & 0 \\ 0 & I \end{bmatrix} + \begin{bmatrix} J_1^* & 0 \\ 0 & I \end{bmatrix}\begin{bmatrix} Y_{11}(t) & Y_{12}(t) \\ Y_{21}(t) & Y_{22}(t) \end{bmatrix}\begin{bmatrix} I & 0 \\ 0 & 0 \end{bmatrix} -$$
$$\left([B_{11}^* B_{21}^*]\begin{bmatrix} Y_{11}(t) & Y_{12}(t) \\ Y_{21}(t) & Y_{22}(t) \end{bmatrix}\begin{bmatrix} I & 0 \\ 0 & 0 \end{bmatrix} + [S_{11}^* S_{21}^*]\right)^* R^{-1}$$
$$\left([B_{11}^* B_{21}^*]\begin{bmatrix} Y_{11}(t) & Y_{12}(t) \\ Y_{21}(t) & Y_{22}(t) \end{bmatrix}\begin{bmatrix} I & 0 \\ 0 & 0 \end{bmatrix} + [S_{11}^* S_{21}^*]\right),$$

(5.35)
$$\begin{bmatrix} I & 0 \\ 0 & 0 \end{bmatrix}\begin{bmatrix} Y_{11}(T) & Y_{12}(T) \\ Y_{21}(T) & Y_{22}(T) \end{bmatrix}\begin{bmatrix} I & 0 \\ 0 & 0 \end{bmatrix} = \begin{bmatrix} M_{11} & M_{12} \\ M_{21} & M_{22} \end{bmatrix}$$

from (5.5), (5.6), where

$$Y(t) = \begin{bmatrix} Y_{11}(t) & Y_{12}(t) \\ Y_{21}(t) & Y_{22}(t) \end{bmatrix} = P_1^{-*} X(t) P_1^{-1}$$

is partitioned analogous to $\begin{bmatrix} J_1 & 0 \\ 0 & I \end{bmatrix}$, and we obtain

(5.36)
$$0 = \begin{bmatrix} I & 0 \\ 0 & 0 \end{bmatrix}\begin{bmatrix} Y_{11} & Y_{12} \\ Y_{21} & Y_{22} \end{bmatrix}\begin{bmatrix} J_1 & 0 \\ 0 & I \end{bmatrix} + \begin{bmatrix} J_1^* & 0 \\ 0 & I \end{bmatrix}\begin{bmatrix} Y_{11} & Y_{12} \\ Y_{21} & Y_{22} \end{bmatrix}\begin{bmatrix} I & 0 \\ 0 & 0 \end{bmatrix}$$
$$+ \begin{bmatrix} Q_{11} & Q_{12} \\ Q_{21} & Q_{22} \end{bmatrix} - \left([B_{11}^* B_{21}^*]\begin{bmatrix} Y_{11} & Y_{12} \\ Y_{21} & Y_{22} \end{bmatrix}\begin{bmatrix} I & 0 \\ 0 & 0 \end{bmatrix} + [S_{11}^* S_{21}^*]\right)^* R^{-1}$$
$$\left([B_{11}^* B_{21}^*]\begin{bmatrix} Y_{11} & Y_{12} \\ Y_{21} & Y_{22} \end{bmatrix}\begin{bmatrix} I & 0 \\ 0 & 0 \end{bmatrix} + [S_{11}^* S_{21}^*]\right)$$

from (5.7).

Using the symmetry of $Y(t)$ the differential equation (5.34) splits into the three equations

(5.37)
$$-\dot{Y}_{11}(t) = Q_{11} + Y_{11}(t)J_1 + J_1^* Y_{11}(t)$$
$$- (Y_{11}(t)B_{11} + Y_{21}(t)B_{21} + S_{11})R^{-1}(B_{11}^* Y_{11}(t) + B_{21}^* Y_{21}(t) + S_{11}^*)$$

(5.38)
$$0 = Q_{21} + Y_{21}(t) - S_{21}R^{-1}(B_{11}^* Y_{11}(t) + B_{21}^* Y_{21}(t) + S_{11}^*)$$

(5.39)
$$0 = Q_{22} - S_{21}R^{-1}S_{21}^*.$$

Equation (5.39) is a consistency condition. If we solve (5.38) for $Y_{21}(t)$, we obtain

(5.40)
$$Y_{21}(t) = (I - S_{21}R^{-1}B_{21}^*)^{-1}(-Q_{21} + S_{21}R^{-1}S_{11}^* + S_{21}R^{-1}B_{11}^* Y_{11}(t)).$$

This can be inserted in (5.37) to yield a standard Riccati equation for $Y_{11}(t)$.

If E is nonsingular, then the discrete analogues to (5.5), (5.6), (5.7) are:

(5.41)
$$E^*X_kE = C^*QC + A^*X_{k+1}A$$
$$- (A^*X_{k+1}B + C^*S)(R + B^*X_{k+1}B)^{-1}(B^*X_{k+1}A + S^*C)$$
for $k = k_0, \ldots, K-1$

and the terminal condition

(5.42)
$$E^*X_KE = C^*MC.$$

The algebraic Riccati equation in this case is

(5.43) $E^*XE = C^*QC + A^*XA - (A^*XB + C^*S)(R + B^*XB)^{-1}(A^*XB + C^*S)^*.$

References for results on solvability of (5.41), (5.42) and (5.43) are for example given in Anderson/Moore [A 6], Arnold [A 9], [A 10], Gohberg et al [G 14], Kwakernaak/Sivan [K 28], Lancaster et al [L 1], [L 2], Sage [S 1], Silverman [S 8]. The standard results are then:

5.44 THEOREM. *Let E be nonsingular, R positive definite, $K < \infty$. Then the optimal solution to problem (1.2) exists, is unique and given by*

(5.45) $u_k = -(R + B^*X_kB)^{-1}(A^*X_kB + C^*S)^*x_k, \quad k = k_0, \ldots, K,$

where X_k satisfies the matrix Riccati difference equation (5.41) with terminal condition (5.42).

PROOF: See Arnold [A 9] or substitute as in Theorem 5.6 and use the proof in Sage [S 1, p. 131]. ∎

We remark here that in general the existence of the inverse in (5.45) is guaranteed if X_k is positive semidefinite. In the infinite time case, $K = \infty$, a summary of results is given in Lancaster/Rodman [L 4]. The result that is most important for us is the case when strong stabilizability and strong detectability is assumed.

5.46 THEOREM. *Let E be nonsingular, R positive definite, (E, A, B) strongly stabilizable and $(E, A, C^*(Q - SR^{-1}S^*)C)$ strongly detectable. Then the optimal control for problem (1.2) with $K = \infty$ exists, is unique and given by the feedback law*

(5.47) $u_k = -(R + R^*XB)^{-1}(A^*XB + C^*S)^*x_k, \quad k = k_0, \ldots, K,$

where X is the unique positive semidefinite solution of (5.43). Furthermore the (closed loop) dynamics of the system obtained with this controls

(5.48) $Ex_{k+1} = [A - (R + B^*XB)^{-1}(A^*XB + S^*C)]x_k$

is asymptotically stable, i.e. $\lim\limits_{k \to \infty} x_k = 0$.

PROOF: After the obvious substitutions as above, the well–known proofs may be applied, e.g. Kwakernaak/Sivan [K 28] or Lee [L 16]. ∎

In the case that E is singular, previous work is by Bender/Laub [B 11], Silverman [S 8].

Again the solution to the optimal control problem (1.2), if it exists is unique if and only if the boundary value problem (3.48), (3.49) has a unique solution. Assume again that $\alpha E - \beta A$ is in Kronecker canonical form and the other matrices are as in the continuous case. Then problem (3.48), (3.49) can be permuted as

$$(5.49) \quad \begin{bmatrix} J_1 & 0 & 0 & 0 & B_{11} \\ Q_{11} & -I & 0 & Q_{12} & S_{11} \\ 0 & 0 & 0 & I & B_{21} \\ Q_{21} & 0 & 0 & Q_{22} & S_{21} \\ S_{11}^* & 0 & 0 & S_{21}^* & R \end{bmatrix} \begin{bmatrix} z_k^{(1)} \\ \mu_k^{(1)} \\ \mu_k^{(2)} \\ z_k^{(2)} \\ u_k \end{bmatrix} = \begin{bmatrix} I & 0 & 0 & 0 & 0 \\ 0 & -J_1^* & 0 & 0 & 0 \\ 0 & 0 & 0 & 0 & 0 \\ 0 & 0 & -I & 0 & 0 \\ 0 & -B_{11}^* & -B_{21}^* & 0 & 0 \end{bmatrix} \begin{bmatrix} z_{k+1}^{(1)} \\ \mu_{k+1}^{(1)} \\ \mu_{k+1}^{(2)} \\ z_{k+1}^{(2)} \\ u_{k+1} \end{bmatrix}$$

$$(5.50) \quad z_{k_0}^{(1)} = z_1^0, \; z_{k_0}^{(2)} = z_k^0, \; \mu_K^{(1)} = M_{11} z_K^{(1)},$$

where u_{k+1} occurs only formally.

To discuss uniqueness, we perform some manipulations with (5.49) and obtain:

$$(5.51) \quad z_{k+1}^{(1)} = J_1 z_k^{(1)} + B_{11} u_k$$

$$(5.52) \quad J_1^* \mu_{k+1}^{(1)} = \mu_k^{(1)} - Q_{11} z_k^{(1)} - (S_{11} - Q_{12} B_{21}) u_k$$

$$(5.53) \quad -B_{11}^* \mu_{k+1}^{(1)} = (S_{11}^* - B_{21}^* Q_{21}) z_k^{(1)} + \tilde{R} u_k,$$

where $\tilde{R} = R - S_{21}^* B_{21} - B_{21}^* S_{21} + B_{21}^* Q_{22} B_{21}$ is as in the continuous case. The general solution for u_k in (5.53) is

$$(5.54) \quad u_k = -\tilde{R}^{-1}[(S_{11}^* - B_{21}^* Q_{21}) z_k^{(1)} + B_{11}^* \mu_{k+1}^{(1)}].$$

We then obtain from (5.51), (5.52), (5.53)

$$(5.55) \quad z_{k+1}^{(1)} = [J_1 - B_{11} \tilde{R}^{-1} (S_{11}^* - B_{21}^* Q_{21})] z_k^{(1)} - B_{11} \tilde{R}^{-1} B_{11}^* \mu_{k+1}^{(1)},$$

$$(5.56) \quad \begin{aligned} & (J_1^* - (S_{11} - Q_{12} B_{21}) \tilde{R}^{-1} B_{11}^*) \mu_{k+1}^{(1)} \\ &= \mu_k^{(1)} - [Q_{11} - (S_{11} - Q_{12} B_{21}) \tilde{R}^{-1} (S_{11}^* - B_{21}^* Q_{21})] z_k^{(1)}, \end{aligned}$$

or in matrix form

$$\begin{bmatrix} I & G \\ 0 & F^* \end{bmatrix} \begin{bmatrix} z_{k+1}^{(1)} \\ \mu_{k+1}^{(1)} \end{bmatrix} = \begin{bmatrix} F & 0 \\ H & I \end{bmatrix} \begin{bmatrix} z_k^{(1)} \\ \mu_k^{(1)} \end{bmatrix}$$

(5.57)

$$= \begin{bmatrix} I & (S_{11} - Q_{12}B_{21})\tilde{R}^{-1} \end{bmatrix} \begin{bmatrix} I & 0 & 0 \\ 0 & -B_{21}^* & I \end{bmatrix} \begin{bmatrix} Q_{11} & Q_{12} & S_{11} \\ Q_{21} & Q_{22} & S_{12} \\ S_{11}^* & S_{21}^* & R \end{bmatrix},$$

where

$$G := B_{11}\tilde{R}^{-1}B_{11}^*, \quad H := -(S_{11} - Q_{12}B_{21})\tilde{R}^{-1}(S_{11}^* - B_{21}^*Q_{21}) + Q_{11},$$
$$F := J_1 - B_{11}\tilde{R}^{-1}(S_{11}^* - B_{21}^*Q_{21}).$$

The general solution of (5.57) is then given by the well-known results for homogeneous difference equations (e.g. Campbell [C 2, p. 68]). To apply these results we need the following Lemma of Mehrmann [M 8]:

5.58 LEMMA.

$$\begin{bmatrix} F & 0 \\ -H & I \end{bmatrix} - \lambda \begin{bmatrix} I & G \\ 0 & F^* \end{bmatrix} = P(\lambda)$$

is a regular pencil.

PROOF: Suppose $P(\lambda)$ is a singular pencil, then for every $\lambda \in \mathbf{C}$ there exists $\begin{bmatrix} w_1 \\ w_2 \end{bmatrix} \in \mathbf{C}^{2n}\backslash\{0\}$ such that

(5.59)
$$\left(\begin{bmatrix} F & 0 \\ -H & I \end{bmatrix} - \lambda \begin{bmatrix} I & G \\ 0 & F^* \end{bmatrix}\right) \begin{bmatrix} w_1 \\ w_2 \end{bmatrix} = 0.$$

Let now $\lambda \in \mathbf{C}$, $|\lambda| = 1$ such that λ is not an eigenvalue of F and let $\begin{bmatrix} w_1 \\ w_2 \end{bmatrix} \neq 0$ be such that (5.59) holds. Then

$$(F - \lambda I)w_1 - \lambda G w_2 = 0, \quad -Hw_1 + (I - \lambda F)w_2 = 0.$$

The same arguments as in Corollary 4.36 show that

(5.60)
$$(|\lambda|^2 - 1)w_2^*w_1 = -(|\lambda|^2 w_2^* G w_2 + w_1^* H w_1).$$

Now the left side of (5.60) is zero. Since G, H are positive semidefinite it follows that $w_1 \in \mathcal{N}(H)$, $w_2 \in \mathcal{N}(G)$. But this is a contradiction, since it implies that $(F - \lambda I)w_1 = 0$ and $(I - \lambda F^*)w_2 = 0$. ∎

Thus, the solution of (5.54) is (e.g. Campbell [C 2])

(5.61)
$$\begin{bmatrix} z_k^{(1)} \\ \mu_k^{(1)} \end{bmatrix} = Z^k \begin{bmatrix} z_{k_0}^{(1)} \\ \mu_{k_0}^{(1)} \end{bmatrix}, \quad k = k_0, \dots, K,$$

where $Z = \left(P(\lambda)^{-1} \begin{bmatrix} I & G \\ 0 & F^* \end{bmatrix}\right)^D \left(P(\lambda)^{-1} \begin{bmatrix} F & 0 \\ -H & I \end{bmatrix}\right)$ for some λ such that $P(\lambda)^{-1}$

exists and an exponent D denotes again the Drazin inverse. Let $Z^K = \begin{bmatrix} Z_{11} & Z_{12} \\ Z_{21} & Z_{22} \end{bmatrix}$. The

boundary conditions (5.49) then yield

$$(5.62) \qquad (M_{11}Z_{11} - Z_{21})z_{k_0}^{(1)} = (-M_{11}Z_{12} + Z_{22})\mu_{k_0}^{(1)},$$

so the solution is unique if and only if $Z_{22} - M_{11}Z_{12}$ is invertible.

We again summarize the results:

5.63 THEOREM. *If $\alpha E - \beta A$ is regular, $\mathrm{ind}_\infty(E, A) \leq 1$ and the boundary conditions (3.49) are consistent, then the solution of the boundary value problem (3.48), (3.49) is unique if and only if $M_{11}Z_{12} - Z_{21}$ is invertible.*

Note that a similar result was already obtained in Bender/Laub [B 11] after some further transformations with the system.

Since $\tilde{R}$ is nonsingular, we again may rewrite equations (5.54), (5.55), (5.56) as

$$(5.64) \qquad \begin{bmatrix} I & 0 & 0 \\ 0 & -J_1^* & 0 \\ 0 & -B_{11}^* & 0 \end{bmatrix} \begin{bmatrix} z_{k+1}^{(1)} \\ \mu_{k+1}^{(1)} \\ u_{k+1} \end{bmatrix} = \begin{bmatrix} J_1 & 0 & B_{11} \\ Q_{11} & -I & S_{11} - Q_{12}B_{21} \\ (S_{11} - Q_{12}B_{21}^*) & 0 & \tilde{R} \end{bmatrix} \begin{bmatrix} z_k^{(1)} \\ \mu_k^{(1)} \\ u_k \end{bmatrix}$$

with boundary conditions

$$(5.65) \qquad z_{k_0}^{(1)} = z_1^0, \quad \mu_K^{(1)} = M_{11}z_K^{(1)}, \quad -B_{21}u_{k_0} = z_2^0.$$

This system may again be viewed as the linear boundary value problem stemming from the control problem.

$$\text{Minimize } \tilde{S}'(z_k^{(1)}, u_k) = \frac{1}{2}(z_K^{(1)*} M_{11} z_K^{(1)}$$

$$(5.66)$$
$$+ \sum_{k=k_0}^{K-1} (z_k^{(1)*} Q_{11} z_k^{(1)} + z_k^{(1)*}(S_{11} - Q_{12}B_{21})u_k + u_k^*(S_{11}^* - B_{21}^* Q_{21})z_k^{(1)} + u_k^* \tilde{R} u_k))$$

subject to the constraints

$$(5.67) \qquad z_{k+1}^{(1)} = J_1 z_k^{(1)} + B_{11}u_k, \quad z_{k_0}^{(1)} = z_1^0, \quad -B_{21}u_{k_0} = x_2^0.$$

Again if (I, J_1, B_{11}) is strongly stabilizable and $(I, J_1, Q_{11} - (S_{11} - Q_{12}B_{21})\tilde{R}^{-1}(S_{11} - Q_{12}B_{21})^*)$ is strongly detectable, then the optimal stabilizing feedback law is

$$(5.68) \qquad u_k = -(\tilde{R} + B_{11}^* X B_{11})^{-1}(J_1^* X B_{11} + S_{11} - Q_{12}B_{21})z_k^{(1)}, \quad k = k_0, \dots, K,$$

where X is the unique positive semidefinite solution of the discrete algebraic Riccati equation

$$(5.69)$$
$$X = Q_{11} + J_1^* X J_1 - (J_1^* X B_{11} + S_{11} - Q_{12}B_{21})(\tilde{R} + B_{11}^* X B_{11})^{-1}(J_1^* X B_{11} + S_{11} - Q_{12}B_{21})^*.$$

The consistency condition is given by

$$(5.70) \qquad [B_{21}(\tilde{R} + B_{11}^* X B_{11})^{-1}(S_{11} - Q_{12}B_{21} + J_1^* X B_{11})]z_1^0 = z_7^0.$$

For $K < \infty$ we get the same condition with X replaced by X_{k_0} .

Asymptotic stability for the complete z_k and hence also for x_k , is again obtained since $z_k^{(2)} = -B_{21}u_k$ is again just a constant matrix multiplied to $z_k^{(1)}$. Thus, if $z_k^{(1)}$ is asymptotically stable, so is $z_k^{(2)}$.

In this section we have shown that under the assumptions of section 3 and under the further assumption that the matrix R is positive definite, we have that solutions are unique if and only if the solution of the system given by the boundary conditions is uniquely solvable. Then under further consistency conditions (5.33), (5.70) the general problems can be reduced to the standard problems, so we get the usual results concerning feedback solutions.

§ 6 Algebraic Riccati equations and deflating subspaces.

In Section 5 we have seen that the optimal stabilizing feedback laws for (1.1), (1.2) are obtainable using the solutions of differential and algebraic Riccati equations. Major tools in the analysis and solution of algebraic Riccati equations are invariant or deflating subspaces of the matrix pencils $\alpha A - \beta B$, $\alpha A' - \beta B'$. Among the many references are Casti [C 4], Emami–Naeini [E 5], Gohberg et al [G 14], Lancaster/Rodman [L 3], Laub [L 12], Paige/Van Loan [P 2], Pappas et al [P 5], Rodman [R 10], Singer/Hammarling [S 10], Van Dooren [V 4] to name only a few. The computation of deflating subspaces of the two pencils, however, can also be used directly to solve the boundary value problems (3.19), (3.20) and (3.48), (3.49) and are also the key methods in the numerical algorithms presented in the following sections.

Let the columns of $\begin{bmatrix} Z_1 \\ Z_2 \\ Z_3 \end{bmatrix} \in C^{2n+m,n}$ span an n–dimensional deflating subspace of $\alpha A - \beta B$, i.e.

$$A \begin{bmatrix} Z_1 \\ Z_2 \\ Z_3 \end{bmatrix} = \begin{bmatrix} W_1 \\ W_2 \\ W_3 \end{bmatrix} U_1, \; B \begin{bmatrix} Z_1 \\ Z_2 \\ Z_3 \end{bmatrix} = \begin{bmatrix} W_1 \\ W_2 \\ W_3 \end{bmatrix} U_2.$$

Having the possible feedback solution in mind we make the approach that the solution $\begin{bmatrix} x(t) \\ \mu(t) \\ u(t) \end{bmatrix}$ is contained in this subspace, and set

(6.1)
$$\begin{bmatrix} x(t) \\ \mu(t) \\ u(t) \end{bmatrix} = \begin{bmatrix} Z_1 \\ Z_2 \\ Z_3 \end{bmatrix} w(t)$$

then we obtain

$$A \begin{bmatrix} x(t) \\ \mu(t) \\ u(t) \end{bmatrix} = A \begin{bmatrix} Z_1 \\ Z_2 \\ Z_3 \end{bmatrix} w(t) = \begin{bmatrix} W_1 \\ W_2 \\ W_3 \end{bmatrix} U_1 w(t)$$

$$B \begin{bmatrix} \dot{x}(t) \\ \dot{\mu}(t) \\ \dot{u}(t) \end{bmatrix} = B \begin{bmatrix} Z_1 \\ Z_2 \\ Z_3 \end{bmatrix} \dot{w}(t) = \begin{bmatrix} W_1 \\ W_2 \\ W_3 \end{bmatrix} U_2 \dot{w}(t).$$

If we then solve the differential equation

(6.2)
$$\begin{bmatrix} W_1 \\ W_2 \\ W_3 \end{bmatrix} U_2 \dot{w}(t) = \begin{bmatrix} W_1 \\ W_2 \\ W_3 \end{bmatrix} U_1 w(t), \; Z_1 w(t_0) = x^0, \; E^* Z_2 w(T) = C^* M C Z_1 w(T),$$

we may use $x(t) = Z_1 w(t)$ to compute the trajectory, and $u(t) = Z_3 w(t)$ to compute the control.

The same ideas can be used to compute the solution in the discrete case.

Now let $\begin{bmatrix} Z_1 \\ Z_2 \\ Z_3 \end{bmatrix}$ be a special deflating subspace, and we are particulary interested in the subspace corresponding to the eigenvalues with negative real part, (or inside the unit circle in the discrete case).

If this is the case, i.e. an n–dimensional deflating subspace $\begin{bmatrix} Z_1 \\ Z_2 \\ Z_3 \end{bmatrix}$ corresponding to the eigenvalues with negative real part exists, then the corresponding pencil

$$\alpha U_1 - \beta U_2$$

has only eigenvalues with negative real part and hence U_2 is invertible and $U_2^{-1}U_1$ is c-stable. Thus, (6.2) has a general solution $w(t) = e^{U_2^{-1}U_1(t-t_0)}q$, and we obtain

(6.3) $$x(t) = Z_1 e^{U_2^{-1}U_1(t-t_0)}q,$$

where q is a solution of $Z_1 q = x^0$,

(6.4) $$u(t) = Z_3 e^{U_2^{-1}U_1(t-t_0)}q,$$

and if Z_1 is invertible, we obtain

(6.5) $$u(t) = Z_3 Z_1^{-1}x(t), \ \mu(t) = Z_2 Z_1^{-1}x(t)$$

as the optimal feedback solution.

Now observe that the feedback solution was, under the above assumptions in the case $T = \infty$, shown to be given via the solution of the algebraic Riccati equation

(6.6) $$0 = C^*QC + E^*XA + A^*XE - (B^*XE + S^*C)^*R^{-1}(B^*XE + S^*C).$$

So it is not surprising that deflating subspaces will also be used to obtain solutions to algebraic Riccati equations, independently of any control problems in the background.

To show this relationship we give the following theorems which are extensions of well–known results, e.g. Lancaster/Rodman [L 3].

6.7 THEOREM. Let $\alpha A - \beta B$ as in (4.1) be a regular pencil and let X be an Hermitian solution of (6.6). Then, the columns of

(6.8) $$V = \begin{bmatrix} I \\ XE \\ -R^{-1}(B^*XE + S^*C) \end{bmatrix},$$

span an n–dimensional subspace of $\mathbb{C}^{2n+m}$, which is a deflating subspace for $\alpha A - \beta B$.

PROOF: Let X be an Hermitian solution of (6.6) and V as in (6.8). Then

(6.9)
$$AV = \begin{bmatrix} A - BR^{-1}(B^*XE + S^*C) \\ C^*QC + A^*XE - C^*SR^{-1}(B^*XE + S^*C) \\ S^*C + B^*XE - B^*XE - S^*C \end{bmatrix} = $$
$$\begin{bmatrix} I \\ -E^*X \\ 0 \end{bmatrix} (A - BR^{-1}(B^*XE + S^*C)),$$

where we have used (6.6) in the second component. For B we obtain

$$\beta V = \begin{bmatrix} E \\ -E^*XE \\ 0 \end{bmatrix} = \begin{bmatrix} I \\ -E^*X \\ 0 \end{bmatrix} E.$$

Thus, the columns of V span a deflating subspace for $\alpha A - \beta B$. ∎

Observe that no assumption on E is required for this result. If E is nonsingular, then the converse of Theorem 6.7 is also well-known (e.g Lancaster/Rodman [L 3]). For the case that E is singular, we have to perform the reductions as in Section 5 to obtain the converse of this result.

We assume again for $\alpha E - \beta A$ is in Kronecker canonical form and perform the reductions as in Section 5, which lead to the reduced problem (5.27), (5.28), (5.29), (5.30) with the corresponding pencil:

$$(6.10) \quad \alpha\hat{A} - \beta\hat{B} := \alpha \begin{bmatrix} J_1 & 0 & B_{11} \\ Q_{11} & J_1^* & S_{11} - Q_{12}B_{21} \\ (S_{11} - Q_{12}B_{21})^* & B_{11}^* & \tilde{R} \end{bmatrix} - \beta \begin{bmatrix} I & 0 & 0 \\ 0 & -I & 0 \\ 0 & 0 & 0 \end{bmatrix}.$$

The corresponding Riccati equation is

$$(6.11) \quad 0 = Q_{11} + \hat{X}J_1 + J_1^*\hat{X} + ((S_{11} - Q_{12}B_{21})^* + B_{11}^*\hat{X})^*\tilde{R}^{-1}(B_{11}^*\hat{X} + (S_{11} - Q_{12}B_{21})^*).$$

We can then apply the following Theorem which is a simple extension of the result in Lancaster/Rodman [L 3].

6.12 THEOREM. *Let E be nonsingular, $\mathcal{R}$ positive definite. Let the columns of*

$$V = \begin{bmatrix} I \\ XE \\ Z \end{bmatrix}$$

span an n-dimensional deflating subspace of $\alpha A - \beta B$.
Then, X is a solution to (6.6).

PROOF: By assumption $\alpha A - \beta B$ is a regular pencil. There exist $U_1, U_2, W_1, W_2 \in \mathbb{C}^{n,n}$, $W_3 \in \mathbb{C}^{m,n}$ such that

$$(6.13) \quad \begin{aligned} \begin{bmatrix} A & 0 & B \\ C^*QC & A^* & C^*S \\ S^*C & B^* & R \end{bmatrix} \begin{bmatrix} I \\ XE \\ Z \end{bmatrix} &= \begin{bmatrix} W_1 \\ W_2 \\ W_3 \end{bmatrix} U_1, \\[2mm] \begin{bmatrix} E & 0 & 0 \\ 0 & -E^* & 0 \\ 0 & 0 & 0 \end{bmatrix} \begin{bmatrix} I \\ XE \\ Z \end{bmatrix} &= \begin{bmatrix} W_1 \\ W_2 \\ W_3 \end{bmatrix} U_2. \end{aligned}$$

$E = W_1 U_2$ nonsingular implies that U_2 is invertible. Thus, by (6.13) we have

$$(6.14) \quad \begin{bmatrix} A & 0 & B \\ C^*QC & A^* & C^*S \\ S^*C & B^* & R \end{bmatrix} \begin{bmatrix} I \\ XE \\ Z \end{bmatrix} = \begin{bmatrix} E & 0 & 0 \\ 0 & -E^* & 0 \\ 0 & 0 & 0 \end{bmatrix} \begin{bmatrix} I \\ XE \\ Z \end{bmatrix} U_2^{-1}U_1.$$

Using the last equation to eliminate Z we obtain with $U = U_2^{-1}U_1$

(6.15) $$A + B(-R^{-1}S^*C - R^{-1}B^*XE) = EU$$

and

(6.16) $$C^*QC + A^*XE + C^*S(R^{-1}S^*C - R^{-1}B^*XE) = -E^*XEU.$$

Substituting (6.15) in (6.16), we obtain

(6.17) $$\begin{aligned} C^*QC + A^*XE - C^*SR^{-1}S^*C - C^*SR^{-1}B^*XE = \\ - E^*XA + E^*XB(R^{-1}S^*C + R^{-1}B^*XE), \end{aligned}$$

which is equivalent to (6.6). ∎

Applying Theorem 6.12 to $\alpha\widehat{A} - \beta\widehat{B}$ of (6.10) we have that X is a solution to (6.11) if

the columns of $\begin{bmatrix} I \\ \widehat{X} \\ Z \end{bmatrix}$ span a deflating subspace of $\alpha\widehat{A} - \beta\widehat{B}$.

For the reduced pencil $\alpha\widehat{A} - \beta\widehat{B}$ we can again apply the whole theory concerning existence and uniqueness of positive semidefinite solutions to (6.11). (E.g. Kučera [K 24], Lancaster/Rodman [L 4].)

For the discrete case again we get analogous results. We study the algebraic Riccati equation

(6.18) $$E^*XE = C^*QC + A^*XA - (A^*XB + C^*S)(R + B^*XB)^{-1}(A^*XB + C^*S)^*.$$

6.19 THEOREM. Let $\alpha A' - \beta B'$ as in (4.2) be a regular pencil and let X be an Hermitian solution of (6.18). Then, the columns of

(6.20) $$V' = \begin{bmatrix} I \\ XE \\ -(R + B^*XB)^{-1}(A^*XB + C^*S)^* \end{bmatrix}$$

span an n–dimensional subspace of $\mathbb{C}^{2n+m}$, which is a deflating subspace for $\alpha A' - \beta B'$.

PROOF: With V' as above we obtain

$$A'V' = \begin{bmatrix} I \\ -A^*X \\ -B^*X \end{bmatrix} [A - B(R + B^*XB)^{-1}(A^*XB + C^*S)^*]$$

$$B'V' = \begin{bmatrix} I \\ -A^*X \\ -B^*X \end{bmatrix} E,$$

and thus the result follows. ∎

For the converse we again have to use the reduced pencil

$$(6.21) \quad \alpha \widehat{A}' - \beta \widehat{B}' = \alpha \begin{bmatrix} J_1 & 0 & B_{11} \\ Q_{11} & -I & S_{11} - Q_{12}B_{21}^* \\ (S_{11} - Q_{12}B_{21}^*)^* & 0 & \widetilde{R} \end{bmatrix} - \beta \begin{bmatrix} I & 0 & 0 \\ 0 & -J_1^* & 0 \\ 0 & -B_{11}^* & 0 \end{bmatrix}$$

and the corresponding Riccati equation

$$(6.22) \quad \begin{aligned} \widehat{X} = Q_{11} + J_1^* \widehat{X} J_1 - \\ (J_1^* \widehat{X} B_{11} + S_{11} - Q_{12}B_{21})(\widetilde{R} + B_{11}^* \widehat{X} B_{11})^{-1}(J_1^* \widehat{X} B_{11} + S_{11} - Q_{12}B_{21})^*. \end{aligned}$$

We can then apply the following result:

(6.23) THEOREM. *Let E be nonsingular, R positive definite. Let the columns of*

$$V' = \begin{bmatrix} I \\ XE \\ Z \end{bmatrix}$$

span an n–dimensional deflating subspace of $\alpha A' - \beta B'$.
Then, X is a solution to (6.18).

PROOF: The proof is completely analogous to the proof of Theorem 6.12. ∎

Clearly if we want Hermitian positive semidefinite solutions to the Riccati equations (6.6), (6.18) we need, as mentioned above, deflating subspaces corresponding to stable eigenvalues, i.e. eigenvalues in the open left half plane or the unit circle, respectively.

In this section we have indicated the relationship between deflating subspaces and solutions to the control problem (1.1), (1.2) and the corresponding algebraic Riccati equations.
This relationship turns out to be very useful for the construction of numerical algorithms, which we will consider in the next sections. But observe that most of the results of sections 3, 4, 5, 6 are theoretical results, which are based on the Kronecker canonical form of a regular pencil. To compute this form numerically is usually an ill–conditioned problem. Thus, for the numerical algorithms, a different reduction has to be used.

§ 7 Schur–forms, Hessenberg–forms and triangular decompositions.

We have seen in Section 5 that a possible way to compute the solutions to the systems of difference or differential equations (3.19), (3.48) and the algebraic Riccati equations (6.6), (0.18), is to compute deflating subspaces for the corresponding matrix pencils $\alpha A - \beta B$, $\alpha A' - \beta B'$.

The basis for most of the numerical algorithms to compute these deflating subspaces are transformations to triangular form or variations of triangular forms. We begin with similarity and equivalence transformations. The first and basic result is the Theorem of Schur [S 3].

7.1 THEOREM [SCHUR–FORM]. *Let $A \in \mathbf{C}^{n,n}$. Then there exists $Q \in \mathcal{U}_n(\mathbf{C})$ such that $Q^*AQ = T$ is upper triangular, with the eigenvalues of A as diagonal elements in any required order.*

The generalization of this theorem for pencils is due to Stewart [S 18] and Moler/Stewart [M 13].

7.2 THEOREM [GENERALIZED SCHUR–FORM]. *Let $A, B \in \mathbf{C}^{n,n}$. Then, there exist $Q, Z \in \mathcal{U}_n(\mathbf{C})$ such that $Q^*AZ = T$ and $Q^*BZ = S$ are both upper triangular and the pairs of diagonal elements (t_{ii}, s_{ii}) can be arranged in any required order.*

Both theorems have their real counterparts:

7.3 THEOREM [REAL SCHUR–FORM]. *Let $A \in \mathbf{R}^{n,n}$. Then, there exists $Q \in \mathcal{U}_n(\mathbf{R})$ such that $T = Q^TAQ$ is quasi upper triangular, again with any required order of the diagonal elements/blocks.*

7.4 THEOREM [GENERALIZED REAL SCHUR–FORM]. *Let $A, B \in \mathbf{R}^{n,n}$. Then, there exist $Q, Z \in \mathcal{U}_n(\mathbf{C})$ such that $Q^TAZ = T$ is quasi upper triangular and $Q^TBZ = S$ is upper triangular. The diagonal blocks can again be arranged in any required order.*

For the proof of Theorems 7.3, 7.4 see Stewart [S 18].

If we have computed the generalized Schur–form $Q^*(\alpha A - \beta B)Z = \alpha T - \beta S$ then let $T[k]$, $S[k]$ denote the k–th leading principal submatrix of T, S respectively and Q_k, Z_k the matrices defined by the first k columns of Q, Z for some k, $1 \le k \le n$. Then $AZ_k = Q_kT[k]$, $BZ_k = Q_kS[k]$, and hence the columns of Z_k span the deflating subspace to the eigenvalues of $\alpha T[k] - \beta S[k]$. Thus, computing the Schur–form allows us to read off the deflating subspaces from the transformation matrices.

Now we come back to our pencils $\alpha A - \beta B$, $\alpha A' - \beta B'$. They have a lot more structure than general pencils as we have seen in Section 4.

If E, R are nonsingular we obtain as in (4.3), (4.4) the equivalent pencils:

(7.5)
$$
\alpha \mathcal{L} - \beta \mathcal{M} =
\alpha \begin{bmatrix} E^{-1}(A - BR^{-1}S^*C) & E^{-1}BR^{-1}B^*E^{-*} & 0 \\ C^*(Q - SR^{-1}S^*)C & -(E^{-1}(A - BR^{-1}S^*C))^* & 0 \\ 0 & 0 & R \end{bmatrix} - \beta \begin{bmatrix} I & 0 & 0 \\ 0 & I & 0 \\ 0 & 0 & 0 \end{bmatrix}
$$

and

(7.6)
$$\alpha \mathcal{L}' - \beta \mathcal{M}' = \\
\alpha \begin{bmatrix} E^{-1}(A - BR^{-1}S^*C) & 0 & 0 \\ C^*(Q - SR^{-1}S^*)C & I & 0 \\ S^*C & 0 & R \end{bmatrix} - \beta \begin{bmatrix} I & -E^{-1}BR^{-1}B^*E^{-*} & 0 \\ 0 & (A^* - C^*SR^{-1}B^*)E^{-*} & 0 \\ 0 & B^* & 0 \end{bmatrix}.$$

In this case, to compute the required deflating subspaces, we can restrict ourselves therefore to eigenvalue problems

(7.7)
$$\mathcal{H} - \lambda I = \begin{bmatrix} F & G \\ H & -F^* \end{bmatrix} - \lambda \begin{bmatrix} I & 0 \\ 0 & I \end{bmatrix}$$

with a Hamiltonian matrix $\mathcal{H}$, or

(7.8)
$$\alpha \widetilde{\mathcal{L}}' - \beta \widetilde{\mathcal{M}}' = \alpha \begin{bmatrix} F & 0 \\ H & I \end{bmatrix} - \beta \begin{bmatrix} I & -G \\ 0 & F^* \end{bmatrix}$$

which is a symplectic pencil, i.e. $\widetilde{\mathcal{L}}' J(\widetilde{\mathcal{L}}')^* = \widetilde{\mathcal{M}}' J(\widetilde{\mathcal{M}}')^*$, (see Table 2.4). F, G, H are the obvious replacements.

If furthermore F is invertible then (7.8) can be transformed to

(7.9)
$$\mathcal{S} - \lambda I = \begin{bmatrix} F + GF^{-*}H & GF^{-*} \\ F^{-*}H & F^{-*} \end{bmatrix} - \lambda I$$

with a symplectic matrix $\mathcal{S}$.

It is obvious that one wishes to find these structures in some way also in the Schur-form.

We now give a list of Theorems which are variations of Theorem (7.1),...,(7.4) for Hamiltonian matrices, symplectic pencils and symplectic matrices. The first two results are due to Paige/Van Loan [P 2].

7.10 THEOREM [HAMILTONIAN SCHUR-FORM]. *Let $\mathcal{H} \in \mathbf{C}^{2n,2n}$ be Hamiltonian and have only eigenvalues with nonzero real part. Then, there exists $Q \in \mathcal{US}_{2n}(\mathbf{C})$ such that*

$$Q^* \mathcal{H} Q = \begin{bmatrix} T & N \\ 0 & -T^* \end{bmatrix}, \quad T, N \in \mathbf{C}^{n,n},$$

where T is upper triangular and $N = N^*$. T can be chosen to have only eigenvalues in the left half plane.

7.11 THEOREM [REAL HAMILTONIAN SCHUR-FORM]. *Let $\mathcal{H} \in \mathbf{R}^{2n,2n}$ be Hamiltonian and have no nonzero pure imaginary eigenvalues. Then, there exists $Q \in \mathcal{US}_{2n}(\mathbf{R})$ such that*

$$Q^T \mathcal{H} Q = \begin{bmatrix} T & N \\ 0 & -T^T \end{bmatrix}, \quad T, N \in \mathbf{R}^{n,n},$$

T quasi upper triangular, $N = N^T$. T can be chosen to have only eigenvalues in the open left half plane or zero.

The following result on the Cayley transformation of symplectic pencils and Hamiltonian matrices, that also will be useful in the sequel, relates the proof of the Hamiltonian Schur-forms to the symplectic Schur-forms.

7.12 THEOREM [CAYLEY TRANSFORMATION]. a) Let $\mathcal{H} \in \mathbf{C}^{2n,2n}$ be a Hamiltonian matrix, then for some λ not in the spectrum of $\mathcal{H}$, we have $(\mathcal{H} + \lambda I)(\mathcal{H} - \lambda I)^{-1} \in S_{2n}(\mathbf{C})$.

b) Let $T - \lambda S$ be a regular symplectic pencil, i.e. $TJT^* = SJS^*$ and $\lambda \in \mathbf{C}$ such that $\det(I - \lambda S) \neq 0$, then $(T - \lambda S)^{-1}(\bar{\lambda}T - S) \in S_{2n}(\mathbf{C})$.

PROOF: For a) see e.g. Byers [B 39] and for b) Mehrmann [M 6]. ∎

We now can easily prove the analogue to Theorem 7.10 for symplectic matrices.

7.13 THEOREM [SYMPLECTIC SCHUR–FORM]. Let $S \in S_{2n}(\mathbf{C})$ have no eigenvalues on the unit circle. Then, there exists $Q \in US_{2n}(\mathbf{C})$ such that

$$Q^*SQ = \begin{bmatrix} T & N \\ 0 & T^{-*} \end{bmatrix}, \quad T, N \in \mathbf{C}^{n,n},$$

T upper triangular. T can be chosen to have all eigenvalues inside the unit circle.

PROOF: The proof can be obtained along the lines of the proof of Theorem 7.10 as given in Paige/Van Loan [P 2]. But it is much simpler to use Theorem 7.12 a). Observe that by using the inverse of the Cayley transformation S can be transformed to the Hamiltonian matrix

$$\mathcal{H} = (S + I)(S - I)^{-1},$$

which exists, since S has no eigenvalues of modulus 1. The eigenvalues of $\mathcal{H}$ then are $\frac{\lambda+1}{\lambda-1}$ for λ an eigenvalue of S and clearly no eigenvalues with real part zero exist. Let Q be the transformation that transformes $\mathcal{H}$ into Hamiltonian Schur–form. Then

$$Q^*\mathcal{H}Q = Q^*(S + I)Q \, Q^*(S - I)^{-1}Q = (Q^*SQ + I)(Q^*SQ - I)^{-1}.$$

Applying again the Cayley transformation yields

(7.14) $\qquad Q^*SQ = (Q^*\mathcal{H}Q + I)(Q^*\mathcal{H}Q - I)^{-1} \triangleq \begin{bmatrix} \searrow & \square \\ 0 & \searrow \end{bmatrix} \begin{bmatrix} \searrow & \square \\ 0 & \searrow \end{bmatrix}^{-1}.$

Thus Q^*SQ is symplectic and has the form $\begin{bmatrix} \searrow & \square \\ 0 & \searrow \end{bmatrix}$, which implies that it has the form $\begin{bmatrix} T & N \\ 0 & T^{-*} \end{bmatrix}$. The second part follows direct from the second part of Theorem (7.10). ∎

Analogously we can prove the theorem for the real case.

7.15 THEOREM [REAL SYMPLECTIC SCHUR–FORM]. Let $S \in S_{2n}(\mathbf{R})$ have no eigenvalues of modulus 1. Then, there exists $Q \in US_{2n}(\mathbf{R})$ such that

$$Q^TSQ = \begin{bmatrix} T & N \\ 0 & T^{-T} \end{bmatrix}, \quad T, N \in \mathbf{R}^{n,n}$$

T quasi upper triangular. T can be chosen to have only eigenvalues inside the unit circle.

In a recent paper Lin [L 22] gives a complete characterization of Hamiltonian and symplectic Schur forms. He shows in a very complicated proof, that these forms exist if and only the algebraic multiplicity of the eigenvalues on the imaginary axis (unit circle, respectively) is even. By our assumptions we always have that no eigenvalues on the imaginary axis (or unit circle respectively) exist. For this reason we do not discuss the general case here.

Now for symplectic pencils as in (7.8) the situation is more complicated. We have different types of theorems depending how "thin" a Schur–form we want to have. To describe those, we need the following Theorem.

7.16 THEOREM [SYMPLECTIC QR–DECOMPOSITION]. *Let* $S \in S_{2n}(\mathbb{C})$. *Then, there exists a unique* $Q \in US_{2n}(\mathbb{C})$ *such that*

$$T = Q^*S = \begin{bmatrix} T_{11} & T_{12} \\ 0 & T_{11}^{-*} \end{bmatrix}, \quad \text{where } T_{11} \in \mathbb{C}^{n,n}$$

is upper triangular and has positive diagonal elements.

PROOF: E.g. Bunse–Gerstner [B 27] or Byers [B 39]. ∎

7.17 THEOREM [GENERALIZED SYMPLECTIC SCHUR–FORM]. *Let* $\alpha T - \beta S$ *be a regular symplectic pencil that has no eigenvalues of modulus 1. Then, there exists* $Q \in U_{2n}(\mathbb{C})$ *and* $Z \in US_{2n}(\mathbb{C})$ *such that*

$$T = Q^*TZ = \begin{bmatrix} T_{11} & T_{12} \\ 0 & T_{22} \end{bmatrix}, \quad S = Q^*SZ = \begin{bmatrix} S_{11} & S_{12} \\ 0 & S_{22} \end{bmatrix}$$

where $T_{11}, S_{11}, T_{22}^*, S_{22}^* \in \mathbb{C}^{n,n}$ *are upper triangular and* $\alpha T - \beta S$ *is a symplectic pencil. Moreover, T, S can be chosen such that* $\alpha T_{11} - \beta S_{11}$ *has only eigenvalues inside the unit circle.*

PROOF: Let T_i, S_i be sequences of nonsingular matrices such that $\alpha T_i - \beta S_i$ are symplectic pencils that have no eigenvalues of modulus 1 and $\lim_{i \to \infty} (\alpha T_i - \beta S_i) = \alpha T - \beta S$.

Then, by applying Theorem 7.13 to the symplectic matrix $T_i^{-1}S_i$ we obtain $Z_i \in US_{2n}(\mathbb{C})$ such that

$$Z_i^*T_i^{-1}S_iZ_i = \begin{bmatrix} T_i & R_i \\ 0 & T_i^{-*} \end{bmatrix}$$

with the required properties. Let $Q_i \in U_{2n}(\mathbb{C})$ such that

$$Q_iS_iZ_i = \begin{bmatrix} S_{11}^{(i)} & S_{12}^{(i)} \\ 0 & S_{22}^{(i)} \end{bmatrix},$$

with $S_{11}^{(i)}$ upper triangular. Such Q_i always exists by the well–known theorem on QR decompositions, e.g. Golub/Van Loan [G 12]. Then clearly

$$Q_i^*T_iZ_i = \begin{bmatrix} T_{11}^{(i)} & T_{12}^{(i)} \\ 0 & T_{22}^{(i)} \end{bmatrix}$$

and

$$Q_i^*S_iZ_i = \begin{bmatrix} S_{11}^{(i)} & S_{12}^{(i)} \\ 0 & S_{22}^{(i)} \end{bmatrix}$$

form a symplectic pencil, since $Z_i \in US_{2n}(\mathbb{C})$.

Using the Bolzano–Weierstraß theorem, it follows that the bounded sequence

$\{(Q_i, Z_i)\}$, $Q_i \in \mathcal{U}_{2n}(\mathbb{C})$, $Z_i \in \mathcal{US}_{2n}(\mathbb{C})$ has a convergent subsequence $\{(Q_{k_i}, Z_{k_i})\}$ and one obtains immediately that the limit $(Q, Z) = \lim_{i \to \infty} (Q_{k_i}, Z_{k_i})$ has $Q \in \mathcal{U}_{2n}(\mathbb{C})$, $Z \in \mathcal{US}_{2n}(\mathbb{C})$ and that Q^*TZ, Q^*SZ have the required form. ∎

Under further stabilizability and detectability conditions, a proof for this theorem is given in Laub [L 12]. Observe that we obtain the same result also in the real case.

Analogously using the real version of Theorem 7.17 and the proof for the generalized real Schur–form of Stewart [S 18] we obtain:

7.18 THEOREM [GENERALIZED REAL SYMPLECTIC SCHUR–FORM]. Let $\alpha T - \beta S$ be a regular real symplectic pencil, that has no eigenvalues of modulus 1. Then, there exist $Q \in \mathcal{U}_{2n}(\mathbb{R})$ and $Z \in \mathcal{US}_{2n}(\mathbb{R})$, such that

$$T = Q^T T Z = \begin{bmatrix} T_{11} & T_{12} \\ 0 & T_{22} \end{bmatrix}$$
$$S = Q^T S Z = \begin{bmatrix} S_{11} & S_{12} \\ 0 & S_{22} \end{bmatrix}$$

where $T_{11}, S_{22}^* \in \mathbb{R}^{n,n}$ are quasi upper triangular, $S_{11}, T_{22}^* \in \mathbb{R}^{n,n}$ are upper triangular and $\alpha T - \beta S$ is a real symplectic pencil. Moreover, S, T can be chosen such that $\alpha T_{11} - \beta S_{11}$ has only eigenvalues inside the unit circle.

If one applies the previous theorems to symplectic pencils of the form

$$\alpha \begin{bmatrix} F & 0 \\ H & I \end{bmatrix} - \beta \begin{bmatrix} I & -G \\ 0 & F^* \end{bmatrix},$$

one is immediately led to the question, whether in Theorems 7.16, 7.18 the blocks T_{22}, S_{11} can be chosen to be identity matrices. In general this is not the case if we require unitary transformations matrices Q, Z.

It is shown in Mehrmann [M 7] that one cannot even produce a form

$$\alpha \begin{bmatrix} T_{11} & T_{12} \\ 0 & T_{22} \end{bmatrix} - \beta \begin{bmatrix} T_{22}^* & S_{12} \\ 0 & T_{11}^* \end{bmatrix}$$

in general with the described transformations.

But if we allow nonunitary transformations from the left, we have:

7.19 THEOREM [GENERALIZED REDUCED SYMPLECTIC SCHUR–FORM].
Let $\alpha T - \beta S$ be a regular symplectic pencil that has no eigenvalues of modulus 1, then there exists a nonsingular $Q \in \mathbb{C}^{2n,2n}$ and $Z \in \mathcal{US}_{2n}(\mathbb{C})$ such that

$$T = Q^*TZ = \begin{bmatrix} T_{11} & T_{12} \\ 0 & I \end{bmatrix}, \quad S = Q^*SZ = \begin{bmatrix} I & S_{12} \\ 0 & T_{11}^* \end{bmatrix}$$

and T, S can be chosen such that T_{11} has only eigenvalues inside the unit circle.

PROOF: Apply Theorem 7.16 to obtain

$$T = \begin{bmatrix} T_{11} & T_{12} \\ 0 & T_{22} \end{bmatrix}, \quad S = \begin{bmatrix} S_{11} & S_{12} \\ 0 & S_{22} \end{bmatrix}$$

as described. Now $\alpha T_{11} - \beta S_{11}$ has all eigenvalues inside the unit circle, thus S_{11}, T_{22} are nonsingular. Multiplying from the left by $\begin{bmatrix} S_{11}^{-1} & 0 \\ 0 & T_{22}^{-1} \end{bmatrix}$ we obtain the required form. Observe that in general the left factor is not unitary. ∎

Similary one obtains a corresponding result in the real case. Note that we always have the restriction that no eigenvalues are on the imaginary axis or on the unit circle, respectively. In the application from optimal control these requirements are met. More general Schur–forms have been proposed by Clements/Glover [C 8], Hammarling/Singer [H 4] and Lin [L 22] also including eigenvalues on the imaginary axis or unit circle.

While the above Schur–forms are the theoretical basis for most of the following algorithms, triangular decompositions as in Theorem 7.17 form the practical basis for the numerical algorithms, which almost all have the same framework as the QR-algorithm of Francis [F 4], [F 5] or the QZ-algorithm of Moler/Stewart [M 13]. A general concept for algorithms of this type is given by Dolla–Dora [D 1], [D 2], Watkins [W 4] and Watkins/Elsner [W 5]. Now the usual QR-, QZ-algorithm do not make use of the structure of $\alpha A - \beta B$, $\alpha A' - \beta B'$, $\mathcal{H} - \lambda I$, $\alpha \tilde{\mathcal{L}}' - \beta \tilde{\mathcal{M}}'$, $S' - \lambda I$ as defined above. Another structural property that the pencil $\alpha A - \beta B$ has is the following: Let $\tilde{\beta} = i\beta$ then the pencil $\alpha A - \tilde{\beta} iB$ is equivalent to the pencil

$$(7.20) \qquad \alpha \begin{bmatrix} 0 & A & B \\ A^* & Q & S \\ B^* & S^* & R \end{bmatrix} - \tilde{\beta} \begin{bmatrix} 0 & iE & 0 \\ -iE & 0 & 0 \\ 0 & 0 & 0 \end{bmatrix} =: \alpha \tilde{A} - \beta \tilde{B}$$

which is an Hermitian indefinite pencil. This property is also destroyed if one applies the usual QZ–algorithm. Algorithms for the computation of deflating subspaces of Hermitian pencils have been discussed in Bunse–Gerstner [B 23] and Bunse–Gerstner/Mehrmann [B 31]. We do not further discuss this topic here.

In order to have algorithms, which preserve the given structures, one has to consider different type of triangular decompositions and different type of initial reductions as defined in Table 2.4.

We summarize the existence results for the different decompositions in the following theorems:

7.21 THEOREM. Let $A \in \mathbf{C}^{m,m}$ be nonsingular.

i) There exists a unique QR-decomposition of A with $Q \in \mathcal{U}_m(\mathbf{C})$, R upper triangular with positive diagonal elements.

ii) Let $m = 2n$, $S \in \mathcal{S}_{2n}(\mathbf{C})$. Then there exists a unique $Q \in \mathcal{US}_{2n}(\mathbf{C})$ such that $S = QR$ is a symplectic QR-decomposition, i.e. $R = \begin{bmatrix} R_{11} & R_{12} \\ 0 & R_{11}^{-*} \end{bmatrix}$, R_{11} upper triangular with positive diagonal elements.

PROOF: i) E.g. Bunse/Bunse–Gerstner [B 23] or Golub/Van Loan [G 12], ii) Theorem 7.16. ∎

In the real case we have:

7.22 THEOREM. *Let $A \in \mathbf{R}^{m,m}$ be nonsingular.*

i) *There exists a unique QR-decomposition of A with $Q \in \mathcal{U}_m(\mathbf{R})$, R upper triangular with positive diagonal elements.*

ii) *Let $m = 2n$. Then there exists an SR-decomposition of A with $S \in \mathcal{S}_{2n}(\mathbf{R})$, $R = \begin{bmatrix} R_{11} & R_{12} \\ R_{21} & R_{22} \end{bmatrix}$ with R_{11}, R_{22}, R_{21} upper triangular and R_{21} has zero diagonal if and only if all leading principal minors of even dimension of $\hat{P}^T J A \hat{P}^T$ are nonzero, where $\hat{P} := [e_1, e_3, \ldots e_{2n-1}, e_2, e_4, \ldots e_{2n}]$.*

iii) *Let $m = 2n$, $A \in \mathcal{S}_{2n}(\mathbf{R})$. Then there exists a unique $Q \in \mathcal{US}_{2n}(\mathbf{R})$ such that $A = QR$, where $R = \begin{bmatrix} R_{11} & R_{12} \\ 0 & R_{11}^{-T} \end{bmatrix}$, R_{11} upper triangular with positive diagonal elements.*

PROOF: i) follows as in the previous Theorem. ii) follows directly from Theorem 11 in Elsner [E 2], e.g. Bunse–Gerstner/Mehrmann [B 32]. iii) See Byers [B 39]. ∎

Uniqueness for SR decompositions is achievable in different ways, see Mehrmann [M 5]. Since we will not make use of these results, we omit them here. The following fairly general result relating the existence of QR-, SR-decompositions to the corresponding existence of transformation to a "Hessenberg type" form is due to Bunse–Gerstner [B 25].

7.23 THEOREM. *Let $A \in \mathbf{K}^{m,m}$ and $X, Y \in \mathbf{K}^{m,m}$ nonsingular and x_1, y_1 the first column of X, Y respectively. (Here $\mathbf{K} = \mathbf{R}$ or $\mathbf{K} = \mathbf{C}$). Let $K(A, x_1, m) = [x_1, Ax_1, A^2 x_1, \ldots, A^{m-1} x_1]$.*

i) *If $K(A, x_1, m)$ is nonsingular and has the decomposition $K(A, x_1, m) = XR$ with R nonsingular upper triangular and X nonsingular, then $H = X^{-1} A X$ is an unreduced upper Hessenberg matrix.*

ii) *If $H = X^{-1} A X$ is an upper Hessenberg marix, then $K(A, x_1, m) = XR$, where R is the upper triangular matrix $R = K(H, e_1, n)$. If H is unreduced then R is nonsingular.*

iii) *If $H = X^{-1} A X$ and $\tilde{H} = Y^{-1} A Y$ are upper triangular with H unreduced, then if x_1, y_1 are linearly dependent it follows that $X^{-1} Y$ is upper triangular.*

This settles the case of QR-decomposition. For the SR-decomposition a slight modification will give a similar result, e.g. Bunse–Gerstner/Mehrmann [B 32] or Bunse–Gerstner [B 25], [B 26].

For Hamiltonian or symplectic matrices it is in general not clear, which is the correct Hessenberg type form. See the remarks in Paige/Van Loan [P 2] or Ammar [A 2]. A general analysis of the existence of transformations to these forms is given in Ammar/Mehrmann [A 4]. Only in the special case of single input or single output systems the forms are clear. They are the Hamiltonian Hessenberg matrices, the symplectic Hessenberg matrices or the symplectic Hessenberg pencils. In these restricted cases the reduction to these forms always exist. See Byers [B 39], [B 41] for the Hamiltonian and Mehrmann [M 6] for the symplectic case. Other special cases of matrices with two or more structures are discussed in Bunse–Gerstner et al [B 28], [B 29].

In this section we have briefly discussed the different Schur–forms, Hessenberg forms and decompositions that we will use in the decomposition algorithms and seen that the well-known Theorem of Schur admits much stronger results if more structure is on hand. We will make use of any underlying structure as much as possible, first in order to create more efficient methods and second in order not to destroy any properties that are inherited from the problem we started with. We review the different forms discussed in this section in the following tables:

63

7.26 TABLE [TRIANGULAR FORMS/SCHUR–FORMS].

type of matrix/pencil	name of form	structure of form
$A \in \mathbf{C}^{n,n}$	upper triangular form Schur–form	
$\alpha A - \beta B \in \mathbf{C}^{n,n}$	upper triangular pencil generalized Schur–form	$\alpha[\,]-\beta[\,]$
$A \in \mathbf{R}^{n,n}$	quasi upper triangular form real Schur–form	diag. blocks of size ≤ 2
$\alpha A - \beta B \in \mathbf{R}^{n,n}$	quasi upper triangular pencil generalized real Schur–form	diag. blocks of size ≤ 2
$A \in \mathbf{R}^{2n,2n}$	J–triangular form	
$A = \begin{bmatrix} A_{11} & A_{12} \\ A_{21} & A_{22} \end{bmatrix} \in \mathcal{S}_{2n}(\mathbf{C})$	symplectic triangular symplectic Schur–form	$A_{11} = A_{22}^{-*}$
$A = \begin{bmatrix} A_{11} & A_{12} \\ A_{21} & A_{22} \end{bmatrix} \in \mathcal{S}_{2n}(\mathbf{R})$	symplectic quasi triangular real symplectic Schur–form	$A_{11} = A_{22}^{-T}$ diagonal blocks of size ≤ 2
$\mathcal{H} = \begin{bmatrix} H_{11} & H_{12} \\ H_{21} & H_{22} \end{bmatrix} \in \mathbf{C}^{2n,2n}$ Hamiltonian	Hamiltonian triangular Hamiltonian Schur–form	$H_{11} = -H_{22}^*$ $H_{12} = H_{12}^*$
$\mathcal{H} = \begin{bmatrix} H_{11} & H_{12} \\ H_{21} & H_{22} \end{bmatrix} \in \mathbf{R}^{2n,2n}$ Hamiltonian	Hamiltonian quasi triangular real Hamiltonian Schur–form	$H_{11} = -H_{22}^*$ $H_{12} = H_{12}^*$ diag. blocks of size ≤ 2

type of matrix/pencil	name of form	structure of form
$\alpha A - \beta B = \alpha \begin{bmatrix} A_{11} & A_{12} \\ A_{21} & A_{22} \end{bmatrix}$ $-\beta \begin{bmatrix} B_{11} & B_{12} \\ B_{21} & B_{22} \end{bmatrix} \in \mathbf{C}^{2n,2n}$ symplectic pencil	symplectic triangular pencil generalized symplectic Schur–form	$A_{11}A_{22}^* = B_{11}B_{22}^*$
$\alpha A - \beta B = \alpha \begin{bmatrix} A_{11} & A_{12} \\ A_{21} & A_{22} \end{bmatrix}$ $-\beta \begin{bmatrix} B_{11} & B_{12} \\ B_{21} & B_{22} \end{bmatrix} \in \mathbf{C}^{2n,2n}$ symplectic pencil	reduced symplectic triangular pencil reduced generalized symplectic Schur–form	$A_{22} = B_{11} = I,\ A_{11} = B_{22}^*$
$\alpha A - \beta B = \alpha \begin{bmatrix} A_{11} & A_{12} \\ A_{21} & A_{22} \end{bmatrix}$ $-\beta \begin{bmatrix} B_{11} & B_{12} \\ B_{21} & B_{22} \end{bmatrix} \in \mathbf{R}^{n,n}$	symplectic quasi triangular pencil generalized real symplectic Schur form	diag. blocks of size ≤ 2 $A_{11}A_{22}^* = B_{11}B_{22}^*$
$\alpha A - \beta B = \alpha \begin{bmatrix} A_{11} & A_{12} \\ A_{21} & A_{22} \end{bmatrix}$ $-\beta \begin{bmatrix} B_{11} & B_{12} \\ B_{12} & B_{22} \end{bmatrix} \in \mathbf{R}^{2n,2n}$ symplectic pencil	reduced symplectic quasi triangular pencil reduced generalized real symplectic Schur– form	diag. blocks of size ≤ 2 $A_{22} = B_{11} = I,\ A_{11} = B_{22}^T$

7.27 TABLE [HESSENBERG FORMS].

type of matrix/pencil	name of Hessenberg form	structure of Hessenberg form
arbitrary $A \in \mathbf{C}^{n,n}$	upper Hessenberg form	
arbitrary pencil $\alpha A - \beta B \in \mathbf{C}^{n,n}$	Hessenberg triangular form	
$\mathcal{H} = \begin{bmatrix} H_{11} & H_{12} \\ H_{21} & H_{22} \end{bmatrix} \in \mathbf{C}^{2n,2n}$, rk $H_{21} = 1$ Hamiltonian	Hamiltonian Hessenberg form	$H_{12} = H_{12}^*,\ H_{11} = -H_{22}^*$ $H_{21} = a e_n e_n^T,\ a \in \mathbf{R}$
$S = \begin{bmatrix} S_{11} & S_{12} \\ S_{21} & S_{22} \end{bmatrix} \in \mathcal{S}_{2n}(\mathbf{C})$, rk $S_{21} = 1$	symplectic Hessenberg form	$S_{21} = p e_n^T$
$\alpha \begin{bmatrix} A_{11} & A_{12} \\ A_{21} & A_{22} \end{bmatrix} - \beta \begin{bmatrix} B_{11} & B_{12} \\ B_{21} & B_{22} \end{bmatrix}$ $\in \mathbf{C}^{2n,2n}$ rk $A_{11} = 1, A_{12} = B_{21} = 0$, $A_{22} = B_{11} = I$	symplectic Hessenberg pencil	
$A \in \mathbf{R}^{2n,2n}$ arbitrary	J–Hessenberg form	
$\mathcal{H} = \begin{bmatrix} H_{11} & H_{12} \\ H_{21} & H_{22} \end{bmatrix} \in \mathbf{R}^{2n,2n}$ Hamiltonian	J–tridiagonal form	$H_{11} = -H_{22}^T$ H_{12} symmetric

§ 8 Perturbation analysis.

The perturbation analysis of the linear quadratic control problem and the related Riccati equations in the case $E = I$ has been the topic of several recent papers, Arnold/Laub [A 12], Byers [B 39], [B 44], Gahinet/Laub [G 3], Kenney/Hewer [K 9], Kenney/Laub [K 10], [K 13], Kenney et al [K 15], Konstantinov et al [K 19], [K 20], Laub [L 8], [L 13], Petkov et al [P 12], [P 13], [P 14].

It is not an easy task to generalize these results to the cases that $E \neq I$ or R is singular, since then also perturbations to $\alpha E - \beta A$ have to be allowed, which may completely change the structure of the problem. An analysis of this general problem is not known and certainly an important topic of further research.

For this reason, we will discuss the perturbation analysis only for the cases $E = I$, $R = I$, $S = 0$. We will show in the following Section how we can preprocess the control problem with unitary transformations in such a way that

(8.1)
$$E = \begin{bmatrix} \Sigma_E & 0 \\ 0 & 0 \end{bmatrix}, A = \begin{bmatrix} A_{11} & A_{12} \\ A_{21} & A_{22} \end{bmatrix}, B = \begin{bmatrix} B_{11} \\ B_{21} \end{bmatrix}, C = [C_{11} \quad C_{12}],$$
$$Q = \begin{bmatrix} Q_{11} & Q_{12} \\ Q_{21} & Q_{22} \end{bmatrix}, M = \begin{bmatrix} M_{11} & M_{12} \\ M_{21} & M_{22} \end{bmatrix}, S = 0, R = I,$$

(E, A, B) strongly stabilizable, (E, A, C) strongly detectable, and Σ_E, A_{22} are nonsingular diagonal matrices.

If R is well conditioned, (we say that a problem is ill–conditioned if small perturbations in the data lead to large perturbations in the solution,) then we set

$$u(t) = -R^{-1}S^*y(t) + v(t).$$

We obtain the new control problem

(8.2)
$$E\dot{x}(t) = (A - R^{-1}S^*C)x(t) + Bv(t)$$
$$y(t) = Cx(t),$$

with cost functional
(8.3)
$$\tilde{S}(x, v) = \frac{1}{2} \left[x^*(T)C^*MCx(T) + \int_{t_0}^{T} x^*(t)(C^*QC - C^*SR^{-1}S^*C)x(t) + v^*(t)Rv(t)dt \right],$$

hence we have already $S = 0$. Then we can perform as shown in the next section a preprocessing such that E, A, B, C, Q, M are as in (8.1).

The inverses of Σ_E, A_{22}, R directly enter the solution formulas for the system, thus clearly if Σ_E, A_{22}, R are ill–conditioned, then the problem is ill–conditioned. Since these condition numbers are directly available they can be used for error estimates. Observe that the condition number of R can be obtained from a condition estimator when performing the Cholesky decomposition of R, e.g. [G 12].

So for simplicity and since it is not known how to treat the general case, we assume that Σ_E, A_{22}, R are well–conditioned. If we partition $x = \begin{bmatrix} x_1 \\ x_2 \end{bmatrix}$ analogous to (8.1), then we can solve for x_2 in the control problem

(8.4)
$$x_2(t) = -A_{22}^{-1}(A_{21}x_1(t) + B_{21}v(t)),$$
$$y(t) = (C_{11} - C_{12}A_{22}^{-1}A_{21})x_1(t) + C_{12}A_{22}^{-1}B_{21}v(t).$$

Inserting this into the problem, we obtain the reduced system

(8.4)
$$\dot{x}_1(t) = \Sigma_E^{-1}[A_{11} - A_{12}A_{22}^{-1}A_{21}]x_1(t) + \Sigma_E^{-1}[B_{11} - A_{12}A_{22}^{-1}B_{21}]v(t)$$
$$=: A_1 x_1(t) + B_1 v(t), \quad x_1(t_0) = x_1^0$$

and a cost functional

(8.5)
$$\tilde{S}(x_1, v) = \frac{1}{2}[x_1^*(T)M_{11}x_1(T)$$
$$+ \int_{t_0}^T (x_1^*(t)(Q_{11} - Q_{12}A_{22}^{-1}A_{21} - (Q_{12}A_{22}^{-1}A_{21})^* + A_{21}^*A_{22}^{-*}Q_{22}A_{22}^{-1}A_{21})x_1(t)$$
$$- x_1^*(t)(Q_{12}A_{22}^{-1}B_{21})v(t) - v^*(t)(Q_{12}A_{22}^{-1}B_{21})^* x_1(t)$$
$$+ v(t)^*(R - B_{21}^*A_{22}^{-*}Q_{22}A_{22}^{-1}B_{21})v(t))dt]$$
$$=: \frac{1}{2}[x_1^*(T)M_1 x_1^*(T) + \int_{t_0}^T [x_1^*(t) \; v(t)] \begin{bmatrix} Q_1 & S_1 \\ S_1^* & R_1 \end{bmatrix} \begin{bmatrix} x_1(t) \\ v(t) \end{bmatrix} dt].$$

Here we have made use of the assumptions on M, see Remark 3.16.
If we perturb the coefficients of the original control problems by small perturbations, then the perturbations in $A_1, B_1, M_1, Q_1, S_1, R_1$ do not have to be small, since the occuring inverses may introduce large errors.
If, however, $\Sigma_E^{-1}, A_{22}^{-1}, R^{-1}$ are well–conditioned, then small perturbations in the coefficients of the original problem will only introduce small perturbations in the coefficients $A_1, B_1, M_1, Q_1, S_1, R_1$, so for the further discussion we may consider problem (8.4), (8.5) with small perturbations in the coefficients $A_1, B_1, M_1, Q_1, S_1, R_1$.
Using the Cholesky decomposition of R_1 we may again remove S_1, R_1 from these equations so that we end up with $S_1 = 0, R_1 = I$ by setting

$$v(t) = -L_1^{-*}(L_1^{-1}S_1^* x_1(t) - v_1(t)),$$

where L_1 is the factor in the Cholesky decomposition of R_1, which clearly is positive definite by assumption, but also has to be assumed to be well-conditioned.
The same reduction as for the continuous case yields for the discrete case the control problem with dynamics

(8.6)
$$x_{k+1}^{(1)} = A_1 x_k^{(1)} + B_1 v_k, \quad x_{k_0}^{(1)} = x_1^0$$

and the cost functional

(8.7)
$$y'(x_1, v) = \frac{1}{2}[x_K^{(1)*}M_1 x_K^{(1)} + \sum_{k=k_0}^{K-1} [x_k^{(1)*} \; v_k] \begin{bmatrix} Q_1 & S_1 \\ S_1^* & R_1 \end{bmatrix} \begin{bmatrix} x_k^{(1)} \\ v_k^* v_k \end{bmatrix}],$$

with $A_1, B_1, Q_1, M_1, S_1, R_1$ as in (8.4), (8.5). Again we may remove S_1, R_1 from these equations.

For these reduced equations one can now apply the standard theory as discussed in Byers [B 44], Kenney/Hewer [K 9], Kenney/Laub [K 10], Konstantinov et al [K 19], [K 20].

The reduced control problems (8.4), (8.5) and (8.6), (8.7) still have the properties that (A_1, B_1) is stabilizable and (A_1, C_1) is detectable in the usual sense, e.g. Casti [C 4] or Knobloch/Kwakernaak [K 18].

Thus, for the two control problems unique stabilizing solutions exist and there also exist unique stabilizing solutions for all $\tilde{A}_1, \tilde{B}_1, \tilde{Q}_1, \tilde{M}_1, \tilde{S}_1, \tilde{R}_1$ in a sufficiently small neighborhood of $A_1, B_1, Q_1, M_1, S_1, R_1$. For the continuous time case, $T = \infty$ this has been shown in Shubert [S 7], in the discrete time case for $K = \infty$, by Konstantinov et al [K 20]. If $T < \infty$, $L < \infty$, this follows by the standard theory for matrix Riccati differential equations, e.g. Reid [R 4].

In view of the preceeding discussion we call control problems (1.1) and (1.2) nearly *ill–posed* if Σ_E, A_{22} in (8.1) are almost singular, or if a small perturbation in the coefficient matrices destroys the properties (E, A, B) strongly stabilizable, (E, A, C) strongly detectable or R positive definite, respectively. Computational problems that are nearly ill–posed are often ill–conditioned, e.g. Demmel [D 4]. The same is true for the discussed control problems, see Arnold [A 9], [A 10], Byers [B 39], [B 44], Kenney/Hewer [K 9], Stewart [S 13], [S 14], [S 15].

Moreover, it is possible to measure the sensitivity of the solution to perturbations in the data, for the case that we discuss here, Byers [B 44], Konstantinov et al [K 19], [K 20]. In the following we discuss only the case $T, K = \infty$, since then we can reduce the perturbation analysis to algebraic Riccati–equations. For the case of Riccati differential equations or differential equations a similar approach can be taken. A detailed discussion on this topic is given in Kenney/Hewer [K 9] and not discussed here.

Thus, in the following we only discuss the algebraic Riccati equations

$$(8.8) \qquad O = Q_1 + XA_1 + A_1^*X - XB_1B_1^*X$$

and in the discrete case

$$(8.9) \qquad \begin{aligned} O &= Q_1 - X + A_1^*XA_1 - A_1^*XB_1(I + B_1^*XB_1)^{-1}B_1^*XA_1 \\ &= Q_1 - X - A_1^*(I + XB_1B_1^*)^{-1}XA_1. \end{aligned}$$

Let $\tilde{X}$ be the unique Hermitian positive semidefinite solution of the Riccati equations (8.8) or (8.9) with data $\tilde{A}_1, \tilde{B}_1, \tilde{Q}_1$ replacing A_1, B_1, Q_1. If $||A_1 - \tilde{A}_1||, ||B_1 - \tilde{B}_1||, ||Q_1 - \tilde{Q}_1||$ are small enough then $\tilde{X}$ is well defined. Here $||\cdot||$ is a vector norm or the corresponding norm on linear operators, e.g. Golub/Van Loan [G 12].

Following Rice [R 7], [R 8] and common usage, Golub/Van Loan [G 12], we define the condition number

$$(8.10) \qquad \kappa(A_1, B_1, Q_1) = \lim_{\substack{\epsilon \to 0 \\ ||A_1 - \tilde{A}_1|| < \epsilon \\ ||B_1 - \tilde{B}_1|| < \epsilon \\ ||Q_1 - \tilde{Q}_1|| < \epsilon}} \frac{||X - \tilde{X}||}{||\tilde{X}||}.$$

As a "rule of thumb", relative errors of magnitude ϵ in the data result in relative errors in solution of magnitude

$$\epsilon\kappa(A_1, B_1, Q_1).$$

If $\kappa(A_1, B_1, Q_1)$ is large, the Riccati equation is ill–conditioned and perturbations in the data (in particular roundoff errors), may cause large changes in the solution. If $\kappa(A_1, B_1, Q_1)$ is small, e.g. approximately one, then perturbation of the data do not cause large changes in the solution.

Explicit approximations of $\kappa(A, B, Q, R)$ in terms of products and quotients of norms are tedious but straightforward to derive.
For example for the continuous Riccati equation (8.8), setting $F = B_1 B_1^*$, it was shown in Byers [B 44] that

$$(8.11) \qquad \frac{1}{4} \kappa(A_1, Q_1, F) \leq \frac{\|\Omega^{-1}\| \, \|Q_1\| + \|\Theta\| \, \|A_1\| + \|\Pi\| \, \|F\|}{\|X\|} \leq 9\kappa(A_1, Q_1, F),$$

where Ω is the linear operator on $\mathbf{C}^{n \times n}$ given by

$$\Omega(Z) = (A_1 - FX)^* Z + Z(A_1 - FX),$$

Θ is the linear operator on $\mathbf{R}^{n \times n}$ given by

$$\Theta(Z) = \Omega^{-1}(Z^T X + XZ)$$

and Π is the linear operator on $\mathbf{R}^{n \times n}$ given by

$$\Pi(Z) = \Omega^{-1}(XZX).$$

Although (8.11) is a complicated expression, it can be explicitly calculated (at least when n and m are small) by standard packaged software, see Byers [B 44], Garbow et al [G 2], Smith et al [S 11]. It is unfortunate that (8.11) depends explicitly on the solution of the Riccati equation but this is unavoidable. The desired stabilizing solution may have a different condition number than one of the other solutions of (8.8). There is no relationship between the magnitude of $\|X\|$ and conditioning. The solution X appears implicitly in Ω, Θ and Π and sometimes cancels the X in the denominator. It is difficult to interpret the meaning of Θ and Π in physical terms. The expression (8.11) shows that the continuous Riccati equation is ill–conditioned when $\|\Omega^{-1}\|$ is large. In terms of an optimal control problem $\|\Omega^{-1}\|$ is large when the optimal closed loop system is "nearly unstable".

Similar but more tedious constructions lead to analogous results for the discrete Riccati equations (8.9), e.g. Konstantinov et al [K 20]. In Kenney/Hewer [K 9] results of Byers [B 44] are sharpened and extended to other norms than the Frobenius norm and there it is shown how the condition number of the Riccati equation and the damping properties of the closed loop system are related.

On the whole, the perturbation analysis of the control problems is still in the process of development. Besides the obvious ill–conditioning, that is due to ill–conditioning of Σ_E, A_{22}, R, the analysis can in principle be reduced to the analysis of algebraic and differential Riccati equations.

§ 9 Numerical preprocessing.

In the theoretical part (Sections 3–8) we have made frequent use of the fact that we can preprocess the system with preliminary feedback to guarantee that

(9.1) $\qquad\qquad \alpha E - \beta(A + BKC)$ is regular and $\text{ind}_\infty(E, A + BKC) \leq 1,$

under the conditions (E, A, B) strongly stabilizable, (E, A, C) strongly detectable. Often we have also made the assumption that E is nonsingular. In some cases we have achieved these properties using the Kronecker canonical form of $\alpha E - \beta A$. Although numerical methods for the computation of the Kronecker structure of a matrix pencils are known, e.g. Demmel/ Kågström [D 5], [D 6], Kågström [K 1], Van Dooren [V 2], since the Kronecker canonical form is very sensitive to perturbations, it is not advisable to use it for the preprocessing in a numerical method.

In this section we first discuss a numerical method due to Bunse–Gerstner et al [B 35] to achieve properties (9.1). These properties can, however, be also achieved by pole placement methods for descriptor systems as described in Fletcher et al [F 3], which are, however, more expensive and have worse numerical stability properties. Following this compression technique we discuss a method of Van Dooren [V 4] that performs a reduction of the pencils (4.1), (4.2) and deflates infinite eigenvalues due to singularities in R.

We begin with the feedback construction that achieves $\alpha E - \beta(A + BKC)$ regular and $\text{ind}_\infty(E, A + BKC) \leq 1$.

9.2 ALGORITHM CREGSVD [REGULARIZATION OF A COMPLEX SYSTEM VIA SINGULAR VALUE DECOMPOSITION].

Given $E, A \in \mathbf{C}^{n,n}$, $B \in \mathbf{C}^{n,m}$, $C \in \mathbf{C}^{p,n}$, (E, A, B) strongly stabilizable, (E, A, C) strongly detectable, this algorithm determines matrices $U, V \in \mathcal{U}_n(\mathbf{C})$, and $K \in \mathbf{C}^{m,p}$ such that

(9.3) $\qquad\qquad\qquad \alpha UEV - \beta U(A + BKC)V$ is regular ,

(9.4) $\qquad\qquad\qquad \text{ind}_\infty(UEV, U(A + BKC)V) \leq 1,$

(9.5) $\qquad\qquad UEV = \begin{bmatrix} \Sigma_E & 0 \\ 0 & 0 \end{bmatrix}$ with $\Sigma_E \in \mathbf{R}^{r,r}$ nonsingular, diagonal ,

(9.6) $\qquad\qquad U(A + BKC)V = \begin{bmatrix} A_{11} & A_{12} \\ A_{21} & A_{22} \end{bmatrix}$ (partitioned as UEV),

with A_{22} nonsingular and if possible

(9.7) $\qquad\qquad\qquad \text{cond}_2(A_{22}) = \min_{\widetilde{K} \in \mathbf{C}^{m,p}} \{\text{cond}_2(\widetilde{A}_{22}(\widetilde{K}))\},$

where $\widetilde{A}_{22}(\widetilde{K})$ is the corresponding block in the partitioned form of

$$U(A + B\widetilde{K}C)V = \begin{bmatrix} \widetilde{A}_{11} & \widetilde{A}_{12} \\ \widetilde{A}_{21} & \widetilde{A}_{22} \end{bmatrix}.$$

Step 1 Let

(9.8)
$$E = U_1 \Sigma_1 V_1^* = U_1 \begin{bmatrix} \Sigma_E & 0 \\ 0 & 0 \end{bmatrix} V_1^*$$

be a singular value decomposition of E, with Σ_E nonsingular, which can like all following singular value decompositions be obtained by the LINPACK DSVDC routine, e.g. Dongarra et al [D 11]. Set

(9.9)
$$c_E := \text{cond}_2(\Sigma_E),$$

(9.10)
$$A_1 := U_1^T A V_1 = \begin{smallmatrix} s_1 \\ n-s_1 \end{smallmatrix} \{ \overbrace{\begin{bmatrix} A_{11} & A_{12} \\ A_{21} & A_{22} \end{bmatrix}}^{s_1 \quad n-s_1},$$

$$B_1 := U_1^T B =: \begin{bmatrix} B_{11} \\ B_{21} \end{bmatrix}, \quad C_1 := C V_1 =: [C_{11} \quad C_{12}],$$

where A_1, B_1, C_1 are partitioned analogous to Σ_1.

Step 2

Let $B_{21} = U_B \begin{bmatrix} \Sigma_B & 0 \\ 0 & 0 \end{bmatrix} V_B^*$ be a singular value decomposition of B_{21}. Transform and partition the matrices as follows

(9.11)
$$A_2 := \begin{bmatrix} I & 0 \\ 0 & U_B^* \end{bmatrix} A_1 = \begin{smallmatrix} s_1 \\ s_2 \\ n-s_1-s_2 \end{smallmatrix} \{ \overbrace{\begin{bmatrix} A_{11} & A_{12} \\ A_{21} & A_{22} \\ A_{31} & A_{32} \end{bmatrix}}^{s_1 \quad n-s_1}$$

$$B_2 = \begin{bmatrix} I & 0 \\ 0 & U_B^* \end{bmatrix} B_1 V_B^* =: \begin{smallmatrix} s_1 \\ s_2 \\ n-s_1-s_2 \end{smallmatrix} \{ \overbrace{\begin{bmatrix} B_{11} & B_{12} \\ \Sigma_B & 0 \\ 0 & 0 \end{bmatrix}}^{s_1 \quad n-s_1}, \quad C_2 := C_1.$$

Step 3

Let $A_{32} = U_2 \begin{bmatrix} \Sigma_2 & 0 \\ 0 & 0 \end{bmatrix} V_2^*$ be the singular value decomposition of A_{32}. Transform and partition as follows

$$A_3 := \begin{bmatrix} I & 0 & 0 \\ 0 & I & 0 \\ 0 & 0 & U_2^* \end{bmatrix} A_2 \begin{bmatrix} I & 0 \\ 0 & V_2 \end{bmatrix} =: \begin{array}{c} {}^{s_1}\{ \\ {}^{s_2}\{ \\ {}^{s_3}\{ \\ {}^{n-s_1-s_2-s_3}\{ \end{array} \overbrace{\begin{bmatrix} A_{11} & A_{12} & A_{13} \\ A_{21} & A_{22} & A_{23} \\ A_{31} & \Sigma_2 & 0 \\ A_{41} & 0 & 0 \end{bmatrix}}^{\begin{array}{ccc} s_1 & s_3 & n-s_1-s_3 \end{array}},$$

(9.12)

$$B_3 := B_2 = \begin{array}{c} {}^{s_1}\{ \\ {}^{s_2}\{ \\ {}^{s_3}\{ \\ {}^{n-s_1-s_2-s_3}\{ \end{array} \overbrace{\begin{bmatrix} B_{11} & B_{12} \\ \Sigma_B & 0 \\ 0 & 0 \\ 0 & 0 \end{bmatrix}}^{\begin{array}{cc} s_2 & m-s_2 \end{array}},$$

$$C_3 := C_2 \begin{bmatrix} I & 0 \\ 0 & V_2 \end{bmatrix} = {}_p\{ \overbrace{\begin{bmatrix} C_{11} & C_{12} & C_{13} \end{bmatrix}}^{\begin{array}{ccc} s_1 & s_3 & n-s_1-s_3 \end{array}}.$$

Step 4

Let $C_{13} = U_C \begin{bmatrix} \Sigma_C & 0 \\ 0 & 0 \end{bmatrix} V_C^*$ be the singular value decomposition of C_{13}. Transform and partition as follows

(9.13)

$$A_4 := A_3 \begin{bmatrix} I & 0 & 0 \\ 0 & I & 0 \\ 0 & 0 & V_C \end{bmatrix} =: \begin{array}{c} {}^{s_1}\{ \\ {}^{s_2}\{ \\ {}^{s_3}\{ \\ {}^{n-s_1-s_2-s_3}\{ \end{array} \overbrace{\begin{bmatrix} A_{11} & A_{12} & A_{13} & A_{14} \\ A_{21} & A_{22} & A_{23} & A_{24} \\ A_{31} & \Sigma_2 & 0 & 0 \\ A_{41} & 0 & 0 & 0 \end{bmatrix}}^{\begin{array}{cccc} s_1 & s_3 & s_4 & n-s_1-s_3-s_4 \end{array}}, \quad B_4 := B_3,$$

$$C_4 := U_C^* C_3 \begin{bmatrix} I & 0 & 0 \\ 0 & I & 0 \\ 0 & 0 & V_C \end{bmatrix} =: \begin{array}{c} {}^{s_4}\{ \\ {}^{p-s_4}\{ \end{array} \overbrace{\begin{bmatrix} C_{11} & C_{12} & \Sigma_C & 0 \\ C_{21} & C_{22} & 0 & 0 \end{bmatrix}}^{\begin{array}{cccc} s_1 & s_3 & s_4 & n-s_1-s_3-s_4 \end{array}},$$

$$E_4 := E_1 = \begin{array}{c} {}^{s_1}\{ \\ {}^{s_2}\{ \\ {}^{s_3}\{ \\ {}^{n-s_1-s_2-s_3}\{ \end{array} \overbrace{\begin{bmatrix} \Sigma_E & 0 & 0 & 0 \\ 0 & 0 & 0 & 0 \\ 0 & 0 & 0 & 0 \\ 0 & 0 & 0 & 0 \end{bmatrix}}^{\begin{array}{cccc} s_1 & s_3 & s_4 & n-s_1-s_3-s_4 \end{array}}.$$

We have the following transformation matrices.

$$(9.14) \qquad U := U_1 \begin{bmatrix} I & 0 \\ 0 & U_B \end{bmatrix} \begin{bmatrix} I & 0 & 0 \\ 0 & I & 0 \\ 0 & 0 & U_2 \end{bmatrix}, \quad V := V_1 \begin{bmatrix} I & 0 \\ 0 & V_2 \end{bmatrix} \begin{bmatrix} I & 0 & 0 \\ 0 & I & 0 \\ 0 & 0 & V_C \end{bmatrix}$$

Since (E, A, B) was strongly stabilizable and (E, A, C) strongly detectable, it follows that (E_4, A_4, B_4) is strongly stabilizable and (E_4, A_4, C_4) strongly detectable, e.g. [M 9], and hence by (2.19), (2.21) we have that $n - s_1 - s_2 - s_3 = n - s_1 - s_3 - s_4 = 0$ and thus, $s_2 = s_4$. On the other hand, by checking s_2, s_3, s_4 we can test strong stabilizabilty and strong detectability. The matrices thus have the following form:

$$(9.15)$$

$$\widehat{A} = \begin{matrix} s_1\{ \\ s_2\{ \\ s_3\{ \end{matrix} \overbrace{\begin{bmatrix} A_{11} & A_{12} & A_{13} \\ A_{21} & A_{22} & A_{23} \\ A_{31} & \Sigma_2 & 0 \end{bmatrix}}^{s_1 \quad s_3 \quad s_2}, \quad \widehat{B} = \begin{matrix} s_1\{ \\ s_2\{ \\ s_3\{ \end{matrix} \overbrace{\begin{bmatrix} B_{11} & B_{12} \\ \Sigma_B & 0 \\ 0 & 0 \end{bmatrix}}^{s_2 \quad m-s_2},$$

$$\widehat{C} = \begin{matrix} s_2\{ \\ p-s_4\{ \end{matrix} \overbrace{\begin{bmatrix} C_{11} & C_{12} & \Sigma_C \\ C_{21} & C_{22} & 0 \end{bmatrix}}^{s_1 \quad s_3 \quad s_2}, \quad \widehat{E} = \begin{matrix} s_1\{ \\ s_2\{ \\ s_3\{ \end{matrix} \overbrace{\begin{bmatrix} \Sigma_E & 0 & 0 \\ 0 & 0 & 0 \\ 0 & 0 & 0 \end{bmatrix}}^{s_1 \quad s_3 \quad s_2}.$$

Let $\Sigma_3 = \gamma I_{s_2}$, where γ is the largest singular value of Σ_2 and set

$$(9.16) \qquad \begin{aligned} K_{11} &= \Sigma_B^{-1}(\Sigma_3 - A_{23})\Sigma_C^{-1}, \\ K_{12} &= -\Sigma_B^{-1}(A_{22} + (\Sigma_3 - A_{23})\Sigma_C^{-1}C_{12})C_{22}^+, \\ K_{21} &= 0, \ K_{22} = 0, \end{aligned}$$

where C_{22}^+ is the Moore-Penrose inverse of C_{22}, e.g. [C 2]. Set

$$(9.17)$$
$$\begin{aligned} c_{A_{22}} &= \mathrm{cond}(\begin{bmatrix} A_{22} & A_{23} \\ \Sigma_2 & 0 \end{bmatrix} + \begin{bmatrix} \Sigma_B & 0 \\ 0 & 0 \end{bmatrix} \begin{bmatrix} K_{11} & K_{12} \\ K_{21} & K_{22} \end{bmatrix} \begin{bmatrix} C_{12} & \Sigma_2 \\ C_{22} & 0 \end{bmatrix}) \\ &= \mathrm{cond}\begin{bmatrix} (A_{22} + (\Sigma_3 - A_{23})\Sigma_C^{-1}C_{12})(I - C_{22}^+C_{22})) & \Sigma_3 \\ \Sigma_2 & 0 \end{bmatrix}. \end{aligned}$$

END

Note that the decisions, which singular values are zero and which are nonzero has to be made numerically. This is certainly one of the most difficult parts of this algorithm. A good heuristic is to set a singular value σ_i to zero if $\sigma_i < c\epsilon\sigma_1$, where c is a constant (typically 1) and ϵ is the machine precision. Note further that if C_{22} is square and nonsingular then $c_{A_{22}} = \mathrm{cond}\,\Sigma_2$.

If we apply the transformations in (9.14) to our original control problem, then we obtain

(9.18)
$$\widehat{x}(t) := V^* x(t), \widehat{x}_k := V^* x_k, \widehat{x}^0 := V^* x^0, \widehat{y}(t) := U_C^* y(t), \widehat{y}_k = U_C^* y_k,$$
$$\widehat{u}(t) := V_B^* u(t), \ \widehat{u}_k := V_B^* u_k, \widehat{M} := U_C^* M U_C, \widehat{Q} = U_C^* Q U_C, \widehat{S} = S U_C$$

(9.19)
$$\widehat{E}\dot{\widehat{x}} = (\widehat{A} + \widehat{B}\widehat{K}\widehat{C})\widehat{x} + \widehat{B}v, \ \widehat{x}(t_0) = \widehat{x}^0, \widehat{y} = \widehat{C}\widehat{x},$$

(9.20)
$$S(\widehat{y}, \widehat{v}) = \frac{1}{2}\left(\widehat{y}^*(T)\widehat{M}\widehat{y}(T) + \int_{t_0}^{T} \widehat{y}^*[\widehat{Q} + \widehat{K}^* R \widehat{K} + \widehat{S}^* \widehat{K} + \widehat{K}^* \widehat{S}]\widehat{y} \right.$$
$$\left. + \widehat{y}^*(\widehat{S} + \widehat{K}^* R)v + v^*(\widehat{S} + \widehat{K}^* R)^*\widehat{y} + v^* Rv]dt \right)$$

in the continuous case and

(9.21)
$$\widehat{E}\widehat{x}_{k+1} = (\widehat{A} + \widehat{B}\widehat{K}\widehat{C})\widehat{x}_k + \widehat{B}v_k, \ \widehat{x}_{k_0} = \widehat{x}^0, \widehat{y}_k = \widehat{C}\widehat{x}_k,$$

(9.22)
$$S'(\widehat{y}_k, v_k) = \frac{1}{2}\left(\widehat{y}_K^* \widehat{M}\widehat{y}_K + \sum_{k=k_0}^{K-1} \left[\widehat{y}_k^*(\widehat{Q} + \widehat{K}^* R \widehat{K} + \widehat{S}^* \widehat{K} + \widehat{K}^* \widehat{S})\widehat{y}_k \right.\right.$$
$$\left.\left. + \widehat{y}_k^*(\widehat{S} + \widehat{K}^* R)v_k + v_k(\widehat{S} + \widehat{K}^* R)^*\widehat{y}_k + v_k^* Rv_k \right] \right)$$

in the discrete case.

If we set

$$E := \widehat{E}, \ A := \widehat{A} + \widehat{B}\widehat{K}\widehat{C}, \ B := \widehat{B}, \ C := \widehat{C},$$
$$M := \widehat{M}, \ Q := \widehat{Q} + \widehat{K}^* R \widehat{K} + \widehat{S}\widehat{K} + \widehat{K}^* \widehat{S}, \ S := \widehat{S} + \widehat{K}^* R$$
$$y(t) := \widehat{y}(t), \ x(t) := \widehat{x}(t), \ u(t) := v(t),$$
$$u_k := v_k, \ y_k := \widehat{y}_k, \ x_k := \widehat{x}_k,$$

then the control problems are of the forms (1.1) or (1.2) but now $\alpha E - \beta A$ is regular, $\mathrm{ind}_\infty(E, A) \leq 1$ and the properties (E, A, B) strongly stabilizable, (E, A, C) strongly detectable are retained. Furthermore we have values for the condition numbers of Σ_E, A_{22}. There are also methods to change the condition number of E as well as the rank of E using output derivative feedback. Such procedures are discussed in Bunse–Gerstner et al [B 35], but we do not discuss them here. In the real case we denote this procedure by REGSVD. We suggest that this procedure preceeds any further computations with control systems of type (1.1), (1.2) if $E \neq I$.

The above procedure also guarantees that E is already in diagonal form. Note that a singular value decomposition of E was already used for a preprocessing by Bender/Laub [B 11], [B 12] and Kunkel/Mehrmann [K 26] under the assumption that $\alpha E - \beta A$ is regular and $\mathrm{ind}_\infty(E, A) \leq 1$.

So we may assume now that $\alpha E - \beta A$ is regular, $\mathrm{ind}_\infty(E, A) \leq 1$,

(9.23)
$$E =: \begin{bmatrix} \Sigma_E & 0 \\ 0 & 0 \end{bmatrix}, \ A =: \begin{bmatrix} A_{11} & A_{12} \\ A_{21} & A_{22} \end{bmatrix} \text{ with } A_{22} \text{ nonsingular},$$

$$B =: \begin{bmatrix} B_{11} \\ B_{21} \end{bmatrix}, \ S^*C =: [\, S_{11} \ \ S_{12} \,], C^*QC =: \begin{bmatrix} Q_{11} & Q_{12} \\ Q_{21} & Q_{22} \end{bmatrix},$$

$$C^*MC = \begin{bmatrix} M_{11} & M_{12} \\ M_{21} & M_{22} \end{bmatrix}$$

Observe that from the form of the matrices in (9.23) we can immediately check the condition on C^*MC, discussed in Remark 3.16.

If we replace the matrices in (9.23) in the matrix pencils (4.1), (4.2) given by the systems (3.19), (3.48), and perform a block permutation, we obtain in the continuous case the system

(9.24)
$$\begin{bmatrix} A_{11} & 0 & A_{12} & 0 & B_{11} \\ Q_{11} & A_{11}^* & Q_{12} & A_{21}^* & S_{11} \\ Q_{21} & A_{12}^* & Q_{22} & A_{22}^* & S_{21} \\ A_{21} & 0 & A_{22} & 0 & B_{21} \\ S_{11}^* & B_{11}^* & S_{21}^* & B_{21}^* & R \end{bmatrix} \begin{bmatrix} x_1 \\ \mu_1 \\ x_2 \\ \mu_2 \\ u \end{bmatrix} = \begin{bmatrix} \Sigma_E & 0 & 0 & 0 & 0 \\ 0 & -\Sigma_E & 0 & 0 & 0 \\ 0 & 0 & 0 & 0 & 0 \\ 0 & 0 & 0 & 0 & 0 \\ 0 & 0 & 0 & 0 & 0 \end{bmatrix} \begin{bmatrix} \dot{x}_1 \\ \dot{\mu}_1 \\ \dot{x}_2 \\ \dot{\mu}_2 \\ \dot{u} \end{bmatrix},$$

$$\begin{bmatrix} x_1(t_0) \\ x_2(t_0) \end{bmatrix} = \begin{bmatrix} x_1^0 \\ x_2^0 \end{bmatrix}, \quad \begin{bmatrix} \Sigma_E & 0 \\ 0 & 0 \end{bmatrix} \begin{bmatrix} \mu_1(T) \\ \mu_2(T) \end{bmatrix} = \begin{bmatrix} M_{11} & M_{12} \\ M_{21} & M_{22} \end{bmatrix} \begin{bmatrix} x_1(T) \\ x_2(T) \end{bmatrix}.$$

If we now set

(9.25)
$$\tilde{A} := A_{11}, \ \tilde{B} := [A_{12} \ \ 0 \ \ B_{11}], \ \tilde{S} := [Q_{12} \ \ A_{21}^* \ \ S_{11}],$$

$$\tilde{Q} := Q_{11}, \ \tilde{M} = M_{11}, \ \tilde{R} = \begin{bmatrix} Q_{22} & A_{22}^* & S_{21} \\ A_{22} & 0 & B_{21} \\ S_{21}^* & B_{21}^* & R \end{bmatrix},$$

$$\tilde{u} := \begin{bmatrix} x_2 \\ \mu_2 \\ u \end{bmatrix}, \ \tilde{x} := x_1, \ \tilde{\mu} := \mu_1,$$

then we have a new system

(9.26)
$$\begin{bmatrix} \tilde{A} & 0 & \tilde{B} \\ \tilde{Q} & \tilde{A}^* & \tilde{S} \\ \tilde{S}^* & \tilde{B}^* & \tilde{R} \end{bmatrix} \begin{bmatrix} \tilde{x} \\ \tilde{\mu} \\ \tilde{u} \end{bmatrix} = \begin{bmatrix} \Sigma_E & 0 & 0 \\ 0 & -\Sigma_E & 0 \\ 0 & 0 & 0 \end{bmatrix} \begin{bmatrix} \dot{\tilde{x}} \\ \dot{\tilde{\mu}} \\ \dot{\tilde{u}} \end{bmatrix}$$

$$\Sigma_E \tilde{\mu}(T) = M_{11}\tilde{x}(T), \quad \tilde{x}(t_0) = x_1^0, \quad [I \ \ 0 \ \ 0]\tilde{u}(t_0) = x_2^0,$$

which has the same form as the original system and thus can be treated in an analogous way. But note that $\tilde{R}$ is not positive definite any more. (See also Section 5.)

In the discrete case we have
(9.27)

$$
\begin{bmatrix}
A_{11} & A_{12} & 0 & 0 & B_{11} \\
A_{21} & A_{22} & 0 & 0 & B_{21} \\
Q_{11} & Q_{12} & -\Sigma_E & 0 & S_{11} \\
Q_{21} & Q_{22} & 0 & 0 & S_{21} \\
S_{11}^* & S_{21}^* & 0 & 0 & R
\end{bmatrix}
\begin{bmatrix}
x_k^{(1)} \\ x_k^{(2)} \\ \mu_k^{(1)} \\ \mu_k^{(2)} \\ u_k
\end{bmatrix}
=
\begin{bmatrix}
\Sigma_E & 0 & 0 & 0 & 0 \\
0 & 0 & 0 & 0 & 0 \\
0 & 0 & -A_{11}^* & -A_{21}^* & 0 \\
0 & 0 & -A_{12}^* & -A_{22}^* & 0 \\
0 & 0 & -B_{11}^* & -B_{21}^* & 0
\end{bmatrix}
\begin{bmatrix}
x_{k+1}^{(1)} \\ x_{k+1}^{(2)} \\ \mu_{k+1}^{(1)} \\ \mu_{k+1}^{(2)} \\ u_{k+1}
\end{bmatrix},
$$

$$
\begin{bmatrix} x_{k_0}^{(1)} \\ x_{k_0}^{(2)} \end{bmatrix} = \begin{bmatrix} x_1^0 \\ x_2^0 \end{bmatrix},
\qquad
\begin{bmatrix} \Sigma_E & 0 \\ 0 & 0 \end{bmatrix}
\begin{bmatrix} \mu_K^{(1)} \\ \mu_K^{(2)} \end{bmatrix}
=
\begin{bmatrix} M_{11} & M_{12} \\ M_{21} & M_{22} \end{bmatrix}
\begin{bmatrix} x_K^{(1)} \\ x_K^{(2)} \end{bmatrix}.
$$

Observe that the component $\mu_k^{(2)}$ does not occur in the system, so exchanging the fourth block column of both sides and defining $\eta_k = -\mu_{k+1}^{(2)}$, we get the new system
(9.28)

$$
\begin{bmatrix}
A_{11} & A_{12} & 0 & 0 & B_{11} \\
A_{21} & A_{22} & 0 & 0 & B_{21} \\
Q_{11} & Q_{12} & -\Sigma_E & A_{21}^* & S_{11} \\
Q_{21} & Q_{22} & 0 & A_{22}^* & S_{21} \\
S_{11}^* & S_{21}^* & 0 & B_{21}^* & R
\end{bmatrix}
\begin{bmatrix}
x_k^{(1)} \\ x_k^{(2)} \\ \mu_k^{(1)} \\ \eta_k \\ u_k
\end{bmatrix}
=
\begin{bmatrix}
\Sigma_E & 0 & 0 & 0 & 0 \\
0 & 0 & 0 & 0 & 0 \\
0 & 0 & -A_{11}^* & 0 & 0 \\
0 & 0 & -A_{12}^* & 0 & 0 \\
0 & 0 & -B_{11}^* & 0 & 0
\end{bmatrix}
\begin{bmatrix}
x_{k+1}^{(1)} \\ x_{k+1}^{(2)} \\ \mu_{k+1}^{(1)} \\ \eta_{k+1} \\ u_{k+1}
\end{bmatrix},
$$

$$
\begin{bmatrix} x_{k_0}^{(1)} \\ x_{k_0}^{(2)} \end{bmatrix} = \begin{bmatrix} x_1^0 \\ x_2^0 \end{bmatrix},
\qquad
\begin{bmatrix} \Sigma_E & 0 \\ 0 & 0 \end{bmatrix}
\begin{bmatrix} \mu_K^{(1)} \\ \eta_K \end{bmatrix}
=
\begin{bmatrix} M_{11} & M_{12} \\ M_{21} & M_{12} \end{bmatrix}
\begin{bmatrix} x_K^{(1)} \\ x_K^{(2)} \end{bmatrix},
$$

where $u_{k+1}, \eta_{k+1}, x_{k+1}$ occur only formally. The same block permutation as in the continuous case yields the system :

$$
(9.29)\quad
\begin{bmatrix}
A_{11} & 0 & A_{12} & 0 & B_{11} \\
Q_{11} & -\Sigma_E & Q_{12} & A_{21}^* & S_{11} \\
Q_{21} & 0 & Q_{22} & A_{22}^* & S_{21} \\
A_{21} & 0 & A_{22} & 0 & B_{21} \\
S_{11}^* & 0 & S_{21}^* & B_{21}^* & 0
\end{bmatrix}
\begin{bmatrix}
x_k^{(1)} \\ \mu_k^{(1)} \\ x_k^{(2)} \\ \eta_k \\ u_k
\end{bmatrix}
=
\begin{bmatrix}
\Sigma_E & 0 & 0 & 0 & 0 \\
0 & -A_{11}^* & 0 & 0 & 0 \\
0 & -A_{12}^* & 0 & 0 & 0 \\
0 & 0 & 0 & 0 & 0 \\
0 & -B_{11}^* & 0 & 0 & 0
\end{bmatrix}
\begin{bmatrix}
x_{k+1}^{(1)} \\ \mu_{k+1}^{(1)} \\ x_{k+1}^{(2)} \\ \eta_{k+1} \\ u_{k+1}
\end{bmatrix}.
$$

With the analogous replacements as in (9.25), we then obtain a system of the form

$$
(9.30)\quad
\begin{bmatrix}
\widetilde{A} & 0 & \widetilde{B} \\
\widetilde{Q} & -\Sigma_E & \widetilde{S} \\
\widetilde{S} & 0 & \widetilde{R}
\end{bmatrix}
\begin{bmatrix}
\widetilde{x}_k \\ \widetilde{\mu}_k \\ \widetilde{u}_k
\end{bmatrix}
=
\begin{bmatrix}
\Sigma_E & 0 & 0 \\
0 & -A_{11}^* & 0 \\
0 & -B_{11}^* & 0
\end{bmatrix}
\begin{bmatrix}
\widetilde{x}_{k+1} \\ \widetilde{\mu}_{k+1} \\ \widetilde{u}_{k+1}
\end{bmatrix},
$$

$$
\Sigma_E \widetilde{\mu}_K = \widetilde{M} \widetilde{x}_K, \quad \widetilde{x}_{k_0} = x_1^0, \quad [I \ \ 0 \ \ 0]\widetilde{u}_{k_0} = x_2^0,
$$

which has the same form as the original system.

If Σ_E is well–conditioned, which we may check as in REGSVD, we may further transform (9.24) and (9.29) to obtain $\Sigma_E = I$. This is done in the following algorithm:

9.31 ALGORITHM. *CINVE [Inverse of E]*

Given boundary value problems (9.24) or (9.29) with Σ_E well–conditioned, this algorithm transforms the system to one with $E = I$.

Set

$$x_k^{(1)} := \Sigma_E x_k^{(1)}, \ A_{11} := A_{11}\Sigma_E^{-1}, \ Q_{11} := \Sigma_E^{-1}Q_{11}\Sigma_E^{-1},$$

$$Q_{21} := Q_{21}\Sigma_E^{-1}, \ A_{21} := A_{21}\Sigma_E^{-1}, \ S_{11} := \Sigma_E^{-1}S_{11}, \ \Sigma_E := I,$$

where the inversion of Σ_E is performed in the obvious way.

END

The real version of this algorithm is called INVE.

In Algorithm 9.2 we have constructed A_{22} to be nonsingular. Thus, if A_{22} is well-conditioned, we may decide to invert A_{22} and reduce the system even further. We get from (9.24) and (9.29) the vectors

$$\mu_2(t) = -A_{22}^{-*}\left[(Q_{21} - Q_{22}A_{22}^{-1}A_{21})x_1(t) + A_{12}^*\mu_1(t) + (S_{21} - Q_{22}A_{22}^{-1}B_{21})u(t)\right],$$

$$\eta_k = -A_{22}^{-*}\left[(Q_{21} - Q_{22}A_{22}^{-1}A_{21})x_k^{(1)} + A_{12}^*\mu_k^{(1)} + (S_{21} - Q_{22}A_{22}^{-1}B_{21})u_k\right],$$

$$x_2(t) = -A_{22}^{-1}[A_{21}x_1(t) + B_{21}u(t)],$$

$$x_k^{(2)} = -A_{22}^{-1}[A_{21}x_k^{(1)} + B_{21}u_k].$$

If we insert this into (9.24) and (9.29) respectively and if we set:

(9.32)
$$A_{11} := A_{11} - A_{12}A_{22}^{-1}A_{21},$$
$$Q_{11} := Q_{11} - Q_{12}A_{22}^{-1}A_{21} - A_{21}^*A_{22}^{-*}Q_{21} + A_{21}^*A_{22}^{-*}Q_{22}A_{22}^{-1}A_{21},$$
$$S_{11} := S_{11} - B_{21}^*A_{22}^{-*}S_{21} - Q_{12}A_{22}^{-1}B_{21} + A_{21}^*A_{22}^{-*}Q_{22}A_{22}^{-1}B_{21},$$
$$B_{11} := B_{11} - A_{12}A_{22}^{-1}B_{21},$$
$$R := R - S_{21}^*A_{22}^{-1}B_{21} - B_{21}^*A_{22}^{-*}S_{21} + B_{21}^*A_{22}^{-*}Q_{22}A_{22}^{-1}B_{21},$$

we obtain the reduced systems

(9.33)
$$\begin{bmatrix} A_{11} & 0 & B_{11} \\ Q_{11} & A_{11}^* & S_{11} \\ S_{11}^* & B_1^* & R \end{bmatrix} \begin{bmatrix} x_1 \\ \mu_1 \\ u \end{bmatrix} = \begin{bmatrix} \Sigma_E & 0 & 0 \\ 0 & -\Sigma_E & 0 \\ 0 & 0 & 0 \end{bmatrix} \begin{bmatrix} \dot{x}_1 \\ \dot{\mu}_1 \\ \dot{u} \end{bmatrix}$$

with boundary conditions

(9.34)
$$x_1(t_0) = x_1^0, \quad \Sigma_E\mu_1(T) = M_{11}x_1(T)$$

and consistency conditions

(9.35)
$$M_{12} = 0, \ M_{22} = 0, \ M_{21} = 0, \ -A_{22}^{-1}A_{21}x_1^0 - A_{22}^{-1}B_{21}u(t_0) = x_2^0$$

and analogously in the discrete case

(9.36)
$$\begin{bmatrix} A_{11} & 0 & B_{11} \\ Q_{11} & -\Sigma_E & S_{11} \\ S_{11}^* & 0 & R \end{bmatrix} \begin{bmatrix} x_k^{(1)} \\ \mu^{(1)} \\ u_k \end{bmatrix} = \begin{bmatrix} \Sigma_E & 0 & 0 \\ 0 & -A_{11}^* & 0 \\ 0 & -B_{11}^* & 0 \end{bmatrix} \begin{bmatrix} x_{k+1}^{(1)} \\ \mu_{k+1}^{(1)} \\ u_{k+1} \end{bmatrix}$$

with boundary conditions

(9.36) $$x_{k_0}^{(1)} = x_1^0, \quad \Sigma_E \mu_K^{(1)} = M_{11} x_K^{(1)}$$

and consistency conditions

(9.38) $$M_{12} = 0, \ M_{22} = 0, \ M_{21} = 0, \ -A_{22}^{-1} A_{21} x_1^0 - A_{22}^{-1} B_{21} u_{k_0} = x_2^0.$$

For the purpose of referencing we denote the algorithm that performs the replacements in (9.32) if A_{22} is well–conditioned by CINVA $_{22}$, respectively INVA $_{22}$ in the real case.

Another assumption that we have used in the theoretical part is that $\mathcal{R} = \begin{bmatrix} C^*QC & S^*C \\ C^*S & R \end{bmatrix}$ is positive definite. If this is the case then it follows that in the replacement in (9.32) we again obtain R positive definite, since

(9.39)
$$R - S_{21}^* A_{22}^{-1} B_{21} - B_{12}^* A_{22}^{-*} S_{21} + B_{21}^* A_{22}^{-*} Q_{22} A_{22}^{-1} B_{21}$$
$$= [I \quad -B_{21}^* A_{22}^{-*}] \begin{bmatrix} R & S_{21} \\ S_{21}^* & Q_{22} \end{bmatrix} \begin{bmatrix} I \\ -A_{22}^{-1} B_{21} \end{bmatrix}.$$

We have that $\begin{bmatrix} R & S_{12} \\ S_{21}^* & Q_{22} \end{bmatrix}$ is a principle submatrix of $\mathcal{R}$ and thus positive definite and hence the new matrix R in (9.32) is positive definite. Observe that for this we only need $\begin{bmatrix} R & S_{12} \\ S_{21}^* & Q_{22} \end{bmatrix}$ positive definite, which is weaker than $\mathcal{R}$ positive definite. Thus, if $\begin{bmatrix} R & S_{12} \\ S_{21}^* & Q_{22} \end{bmatrix}$ was positive definite, then R in (9.32), (9.36) is positive definite and we can reduce the systems even further by the following procedure

9.40 ALGORITHM CINVRL [INVERSION OF R].

Given systems of the forms (9.33), (9.36) and R well–conditioned, this algorithms solves for u, u_k in (9.33), (9.36) and yields a condition estimate for R.

Step 1

> Perform a Cholesky decomposition $R = LL^*$ via the algorithm DPOCO in LINPACK, and determine a condition estimate c_R for R.

Step 2 Compute

(9.41) $$F := A_{11} - B_{11} R^{-1} S_{11}^*, \ H := Q_{11} - S_{11} R^{-1} S_{11}^*, \ G = B_{11} R^{-1} B_{11}^*.$$

> by using the factorization of Step 1 and solving linear systems.

END

The real version of this method is called INVRL.

If the system has been rewritten as in (9.26) or (9.30), we may also invert $\widetilde{R}$ to reduce the system, but since $\widetilde{R}$ is not positive definite any more, we use in Step 1 of Algorithm

9.40 a decomposition $R = LDL^*$, with D blockdiagonal, with blocks of size ≤ 2. This decomposition can be obtained via the algorithm DSICO in LINPACK. Using this decomposition, we can then invert R as in Algoritm 9.40. In this case we call the algorithm CINVRB or INVRD in the real case.

After Algorithm 9.40, or any of the discussed variations of it, has been performed, system (9.33) or (9.26) has the form

$$(9.42) \qquad \begin{bmatrix} F & -G \\ H & F^* \end{bmatrix} \begin{bmatrix} x_1(t) \\ \mu_1(t) \end{bmatrix} = \begin{bmatrix} \Sigma_E & 0 \\ 0 & -\Sigma_E \end{bmatrix} \begin{bmatrix} \dot{x}_1(t) \\ \dot{\mu}_1(t) \end{bmatrix}$$

and system (9.36) or (9.30) has the form

$$(9.43) \qquad \begin{bmatrix} F & 0 \\ H & -\Sigma_E \end{bmatrix} \begin{bmatrix} x_k^{(1)} \\ \mu_k^{(1)} \end{bmatrix} = \begin{bmatrix} \Sigma_E & G \\ 0 & -F^* \end{bmatrix} \begin{bmatrix} x_{k+1}^{(1)} \\ \mu_{k+1}^{(1)} \end{bmatrix} .$$

Algorithms CINVE, CINVA $_{22}$, CINVRL, CINVRB (or their real counterparts) can be performed independently of each other and condition numbers are obtained. Algorithm 9.40 provides a condition estimate for R, so all the reduction can be performed in a numerically reliable way, while monitoring the condition numbers. If any of these three condition numbers is large, it follows that the problem itself is ill–conditioned. A better conditioning of the whole problem may be obtained by using preliminary derivative feedbacks, e.g. Bunse–Gerstner et al [B 35], Chapman et al [C 6]. This is still a topic under investigation.

If $\mathcal{R} = \begin{bmatrix} C^*QC & S^*C \\ C^*S & R \end{bmatrix}$ or R is singular then the theory of the problem is more difficult. This case has been studied in detail by Geerts [G 7], [G 8], [G 9], [G 10] and Trentelman [T 4]. A compression technique for this case was discussed in Van Dooren [V 4]. It is described in the following algorithm.

9.44 ALGORITHM CVDCOMP [VAN DOOREN'S COMPRESSION TECHNIQUE IF R IS SINGULAR OR ILL–CONDITIONED].

Given systems of the form (9.33) or (9.36) and assume that $\begin{bmatrix} B_{11} \\ S_{11} \\ R \end{bmatrix} \in C^{2n+m,m}$ has full rank. This algorithm reduces the system to one in which the control variable does not occur explicitly.

Determine $U \in U_{2n+m}(\mathbf{C})$ such that

$$U = \begin{matrix} 2n\{ \\ m\{ \end{matrix} \overbrace{\begin{bmatrix} U_{11} \\ U_{21} \end{bmatrix}}^{2n} \overbrace{\begin{bmatrix} U_{12} \\ U_{22} \end{bmatrix}}^{m} \text{ and } U \begin{bmatrix} B_{11} \\ S_{11} \\ R \end{bmatrix} = \begin{bmatrix} 0 \\ 0 \\ R_1 \end{bmatrix}$$

using the QR-decomposition, e.g. Golub/Van Loan [G 12]. Set

$$U \begin{bmatrix} A_{11} & 0 & B_{11} \\ Q_{11} & A_{11}^* & S_{11} \\ S_{11}^* & B_{11}^* & R \end{bmatrix} =: \begin{bmatrix} M_{11} & M_{12} & 0 \\ M_{21} & M_{22} & 0 \\ M_{31} & M_{32} & R_1 \end{bmatrix},$$

$$U \begin{bmatrix} A_{11} & 0 & B_{11} \\ Q_{11} & -\Sigma_E & S_{11} \\ S_{11}^* & 0 & R \end{bmatrix} =: \begin{bmatrix} M_{11}' & M_{12}' & 0 \\ M_{21}' & M_{22}' & 0 \\ M_{31}' & M_{32}' & R_1 \end{bmatrix}$$

(9.45)

$$U \begin{bmatrix} \Sigma_E & 0 & 0 \\ 0 & -\Sigma_E & 0 \\ 0 & 0 & 0 \end{bmatrix} =: \begin{bmatrix} N_{11} & N_{12} & 0 \\ N_{21} & N_{22} & 0 \\ N_{31} & N_{32} & 0 \end{bmatrix},$$

$$U \begin{bmatrix} \Sigma_E & 0 & 0 \\ 0 & -A_{11}^* & 0 \\ 0 & -B_{11}^* & 0 \end{bmatrix} =: \begin{bmatrix} N_{11}' & N_{12}' & 0 \\ N_{21}' & N_{22}' & 0 \\ N_{31}' & N_{32}' & 0 \end{bmatrix}.$$

Then we have the new systems

(9.46)
$$\begin{bmatrix} M_{11} & M_{12} & 0 \\ M_{21} & M_{22} & 0 \\ M_{31} & M_{32} & R_1 \end{bmatrix} \begin{bmatrix} x_1 \\ \mu_1 \\ u \end{bmatrix} = \begin{bmatrix} N_{11} & N_{12} & 0 \\ N_{21} & N_{22} & 0 \\ N_{31} & N_{32} & 0 \end{bmatrix} \begin{bmatrix} \dot{x}_1 \\ \dot{x}_2 \\ \dot{u} \end{bmatrix},$$

(9.47)
$$\begin{bmatrix} M_{11}' & M_{12}' & 0 \\ M_{21}' & M_{22}' & 0 \\ M_{31}' & M_{32}' & R_1 \end{bmatrix} \begin{bmatrix} x_k^{(1)} \\ \mu_k^{(1)} \\ u_k \end{bmatrix} = \begin{bmatrix} N_{11}' & N_{12}' & 0 \\ N_{21}' & N_{22}' & 0 \\ N_{31}' & N_{32}' & 0 \end{bmatrix} \begin{bmatrix} x_{k+1}^{(1)} \\ \mu_{k+1}^{(1)} \\ u_{k+1} \end{bmatrix}$$

and, since R_1 is nonsingular, we can solve for u, u_k to obtain

(9.48)
$$u(t) = R_1^{-1}\left(-M_{31}x_1(t) - M_{32}\mu_1^{(1)}(t) + N_{31}x_1(t) + N_{32}\mu_1(t)\right),$$
$$u_k = R_1^{-1}\left(-M_{21}'x_k^{(1)} - M_{32}'\mu_k^{(1)} + N_{31}'x_{k+1}^{(1)} + N_{32}'\mu_{k+1}^{(1)}\right),$$

where $x_1(t), \mu_1(t), x_k^{(1)}, \mu_k^{(1)}$ are given by the systems

(9.49)
$$\begin{bmatrix} M_{11} & M_{12} \\ M_{21} & M_{22} \end{bmatrix} \begin{bmatrix} x_1 \\ \mu_1 \end{bmatrix} = \begin{bmatrix} N_{11} & N_{12} \\ N_{21} & N_{22} \end{bmatrix} \begin{bmatrix} \dot{x}_1 \\ \dot{\mu}_1 \end{bmatrix},$$

(9.50)
$$\begin{bmatrix} M_{11}' & M_{12}' \\ M_{21}' & M_{22}' \end{bmatrix} \begin{bmatrix} x_k^{(1)} \\ \mu_k^{(1)} \end{bmatrix} = \begin{bmatrix} N_{11}' & N_{12}' \\ N_{21}' & N_{22}' \end{bmatrix} \begin{bmatrix} x_{k+1}^{(1)} \\ \mu_{k+1}^{(1)} \end{bmatrix},$$

respectively.

END

Unfortunately this compression destroys the structure of (9.33), (9.36) completely but at this time no other structure preserving reduction method for the case that R is singular is known.

In this section we have shown how the control problems can be reduced in a numerically reliable way to boundary value problems with pencils of the form (4.5) or (4.6) with $E = I$, i.e. Hamiltonian or symplectic pencils. We have obtained tests for the conditioning of this reduction via the condition numbers of

$$\Sigma_E, A_{22}, R.$$

Together with these condition estimates a reliable and efficient reduction can be carried out numerically.

§ 10 Defect correction.

In the following sections we will discuss methods for the numerical solution of the algebraic Riccati equation. All these methods are iterative methods and, since they can only be implemented in finite arithmetic, they are subject to roundoff errors. Also some of the methods that we discuss are not numerically backwards stable. For these reasons it is important to have a method for the iterative refinement of the solution. For algebraic Riccati equations such methods can be obtained very nicely using a defect correction approach. We discuss the theoretical basis for this approach in this section and describe a general defect correction method due to Mehrmann/Tan [M 10]. Since most of the methods that we discuss work with deflating subspaces, also methods for the iterative refinement of the subspaces could be used to improve the results, e.g. Demmel [D 3] or Dongarra et al [D 12].

As we have seen in Section 5 we have to compute the Hermitian positive semidefinite solution of the algebraic Riccati equation

$$
\begin{aligned}
(10.1) \quad 0 &= C^*QC + E^*XA + A^*XE - (B^*XE + S^*C)^*R^{-1}(B^*XE + S^*C) \\
&= C^*(Q - SR^{-1}S^*)C + E^*X(A - BR^{-1}S^*C) + (A^* - C^*SR^{-1}B^*)XE \\
&\quad - E^*XBR^{-1}B^*XE.
\end{aligned}
$$

in the continuous case and the Hermitian positive semidefinite solution of

$$
(10.2) \quad E^*XE = C^*QC + A^*XA - (A^*XB + C^*S)(R + B^*XB)^{-1}(A^*XB + C^*S)^*
$$

in the discrete case.

Defect correction for these equations is based on the following two theorems of Mehrmann/Tan [M 10].

10.3 THEOREM. *Let* X *be an Hermitian solution of (10.1) and let* $\tilde{X}$ *be an Hermitian approximation to this solution. Let* $P = X - \tilde{X}$, $\tilde{A} = A - BR^{-1}(B^*\tilde{X}E + S^*C)$ *and let* $\tilde{Q}$ *be the residual obtained by inserting* $\tilde{X}$ *in (10.1), i.e.*

$$
\tilde{Q} = C^*QC + E^*\tilde{X}A + A^*\tilde{X}E - (B^*\tilde{X}E + S^*C)^*R^{-1}(B^*\tilde{X}E + S^*C),
$$

then P *satisfies the algebraic Riccati equation*

$$
(10.4) \quad 0 = \tilde{Q} + E^*P\tilde{A} + \tilde{A}^*PE - E^*PBR^{-1}B^*PE.
$$

PROOF:

$$
\begin{aligned}
0 &= A^*(\tilde{X} + P)E + E^*(\tilde{X} + P)A \\
&\quad + C^*QC - (B^*(\tilde{X} + P)E + S^*C)^*R^{-1}(B^*(\tilde{X} + P)E + S^*C) \\
&= A^*\tilde{X}E + A^*PE + E^*\tilde{X}A + E^*PA + C^*QC - (B^*\tilde{X}E + S^*C)^*R^{-1}(B^*\tilde{X}E + S^*C) \\
&\quad - (B^*\tilde{X}E + S^*C)^*R^{-1}B^*PE - (B^*PE)^*R^{-1}(B^*\tilde{X}E + S^*C) - (B^*PE)^*R^{-1}B^*PE \\
&= \tilde{Q} + E^*P\tilde{A} + \tilde{A}^*PE - E^*PBR^{-1}B^*PE.
\end{aligned}
$$

■

If $S = 0$, as is the case in many practical examples, then (10.1) and (10.4) are exactly of the same type.

In the discrete case we get

10.5 THEOREM. *Let* X *be an Hermitian solution of (10.2) and let* $\hat{X}$ *be an Hermitian approximation to* X. *Let* $V = X - \hat{X}$. *If* $\hat{R} = R + B^*\hat{X}B$ *and* $I + VB\hat{R}^{-1}B^*$ *are nonsingular, then* V *satisfies the equation*

$$(10.6) \qquad 0 = \hat{Q} - E^*VE + \hat{A}^*V\hat{A} - (\hat{A}^*VB)(\hat{R} + B^*VB)^{-1}(\hat{A}^*VB)^*$$

where

$$\hat{A} = -B\hat{R}^{-1}S^*C + (I - B\hat{R}^{-1}B^*\hat{X})A$$

and $\hat{Q}$ *is the residual obtained by inserting* $\hat{X}$ *in (10.2), i.e.*

$$\hat{Q} = -E^*\hat{X}E + C^*QC + A^*\hat{X}A - (A^*\hat{X}B + C^*S)(R + B^*\hat{X}B)^{-1}(A^*\hat{X}B + C^*S)^*.$$

PROOF: Replace X by $\hat{X} + V$ in (10.2) and set $\Lambda = B\hat{R}^{-1}B^*$, $\hat{S} = C^*S\hat{R}^{-1}B^*$. Then we apply the Sherman–Morrison–Woodbury formula, e.g. Golub/Van Loan [G 12] and obtain

$$(10.7) \qquad (\hat{R} + B^*VB)^{-1} = \hat{R}^{-1} - \hat{R}^{-1}B^*(I + VB\hat{R}^{-1}B^*)^{-1}VB\hat{R}^{-1}$$

$$(10.8) \qquad VB(\hat{R} + B^*VB)^{-1}S^*C = V(I + \Lambda V)^{-1}\hat{S}^*$$

$$(10.9) \qquad B(\hat{R} + B^*VB)^{-1}B^*V = I - (I + \Lambda V)^{-1} = \Lambda(I + V\Lambda)^{-1}V.$$

Then we obtain from (10.2):

$$\begin{aligned}
&E^*\hat{X}E - A^*\hat{X}A + (A^*\hat{X}B + C^*S)\hat{R}^{-1}(A^*\hat{X}B + C^*S)^* - C^*QC \\
&+ E^*VE = A^*VA - A^*VB(\hat{R} + B^*VB)^{-1}(A^*VB)^* \\
&+ (A^*\hat{X}B + C^*S)(\hat{R}^{-1}B^*(I + V\Lambda)^{-1}VB\hat{R}^{-1})(A^*\hat{X}B + C^*S)^* \\
&- A^*VB(\hat{R} + B^*VB)^{-1}(A^*\hat{X}B + C^*S)^* - (A^*\hat{X}B + C^*S)(\hat{R} + B^*VB)^{-1}(A^*VB)^* \\
&= A^*VA + A\hat{X}\Lambda(I + V\Lambda)^{-1}V\Lambda\hat{X}A + \hat{S}(I + V\Lambda)^{-1}V\Lambda\hat{X}A \\
&+ A^*\hat{X}\Lambda(I + V\Lambda)^{-1}V\hat{S}^* + \hat{S}(I + V\Lambda)^{-1}V\hat{S}^* - A^*(I - (I + V\Lambda)^{-1})\hat{X}A \\
&- A^*VB(I + \Lambda V)^{-1}\hat{S}^* - A^*\hat{X}(I - (I + \Lambda V)^{-1})A \\
&- \hat{S}(I + \Lambda V)^{-1}B^*VA - A^*V(I - (I + \Lambda V)^{-1})A.
\end{aligned}$$

Setting $\hat{A} = A - \Lambda\hat{X}A - \hat{S}^*$ and $\hat{Q}$ as above, we obtain

$$E^*VE = \hat{Q} + \hat{A}^*V\hat{A} - (\hat{A}^*VB)(\hat{R} + B^*VB)^{-1}(\hat{A}^*VB)^*,$$

which is equivalent to (10.6). ∎

Again observe that for $S = 0$ (10.2) and (10.6) have the same form.

If follows that any method that can be used to solve (10.1) or (10.2) also serves as a method for the solution of (10.4) or (10.6) respectively. But (10.4), (10.6) have many solutions and since we are interested in the (unique) Hermitian positive semidefinite solution of (10.1), (10.2), we have to decide for the correct solution of (10.4) or (10.6). To do this we have the following results.

10.10 LEMMA. *Given the algebraic Riccati equation (10.1) (or (10.2)) and assume that* (E, A, B) *is strongly stabilizable. Then for* $\tilde{A}$ *as in Theorem 10.3 (or* $\hat{A}$ *as in Theorem 10.5) we have* $(E, \tilde{A}, B)$ *strongly stabilizable* ($(E, \hat{A}, B)$ *strongly stabilizable) respectively.*

PROOF: We have
$$\tilde{A} = A - BR^{-1}(B^*\tilde{X}E + S^*C) = A - B\tilde{F}$$
and
$$\hat{A} = (I - B\hat{R}^{-1}B^*\hat{X})A - B\hat{R}^{-1}S^*C = A - B\hat{F}.$$

In both cases this is a standard feedback of the system, which leaves strongly stabilizable systems, strongly stabilizable, e.g. Mehrmann/Krause [M 9]. ∎

In the following we will discuss only the case that E is nonsingular. We have already seen how a reduction to such a system can be done with the numerical techniques in Section 9.

The solution of (10.1) can be obtained via the deflating subspace corresponding to the stable eigenvalues (having negative real part) of the pencil

$$(10.11) \qquad \begin{bmatrix} A - BR^{-1}S^*C & BR^{-1}B^* \\ C^*(Q - SR^{-1}S^*)C & -(A - BR^{-1}S^*C)^* \end{bmatrix} - \lambda \begin{bmatrix} E & 0 \\ 0 & E^* \end{bmatrix}.$$

By Corollary 4.38 it follows that this pencil has exactly n stable eigenvalues and by Lemma 10.10 the same holds for the pencil

$$(10.12) \qquad \begin{bmatrix} \tilde{A} & BR^{-1}B^* \\ \tilde{Q} & -\tilde{A}^* \end{bmatrix} - \lambda \begin{bmatrix} E & 0 \\ 0 & E^* \end{bmatrix}.$$

Let the columns of $\begin{bmatrix} Z_1 \\ Z_2 \end{bmatrix} \in \mathbb{C}^{2n,n}$ span this stable deflating subspace, i.e.

$$(10.13) \qquad \begin{bmatrix} \tilde{A} & BR^{-1}B^* \\ \tilde{Q} & -\tilde{A}^* \end{bmatrix} \begin{bmatrix} Z_1 \\ Z_2 \end{bmatrix} = \begin{bmatrix} E & 0 \\ 0 & E^* \end{bmatrix} \begin{bmatrix} Z_1 \\ Z_2 \end{bmatrix} Z$$

with $Z \in \mathbb{C}^{n,n}$ having only eigenvalues with negative real part.

Since by Lemma 10.10 we have that $(E, \tilde{A}, B)$ is strongly stabilizable, it follows that $(E^{-1}\tilde{A}, E^{-1}B)$ is stabilizable in the usual sense, e.g. Casti [C 4]. Thus, we can apply the standard theory to show that Z_1 is invertible, e.g. Potter [P 16], Kucera [K 23].

Let $P = -Z_2 Z_1^{-1}$ we then have that

$$(10.14) \qquad \begin{bmatrix} \tilde{A} & BR^{-1}B^* \\ \tilde{Q} & -\tilde{A}^* \end{bmatrix} \begin{bmatrix} I \\ -P \end{bmatrix} = \begin{bmatrix} E & 0 \\ 0 & E^* \end{bmatrix} \begin{bmatrix} I \\ -P \end{bmatrix} \tilde{Z},$$

where $\tilde{Z} = Z_1 Z Z_1^{-1}$ is also c–stable. From (10.14) we obtain

$$\begin{bmatrix} E^{-1}\tilde{A} & E^{-1}BR^{-1}B^*E^{-*} \\ \tilde{Q} & -\tilde{A}^*E^{-*} \end{bmatrix} \begin{bmatrix} I \\ -E^*P \end{bmatrix} = \begin{bmatrix} I \\ -E^*P \end{bmatrix} \tilde{Z}.$$

Thus, by the discussion in Section 6 we get that

(10.15) $$E^{-1}\tilde{A} - E^{-1}BR^{-1}B^*E^{-*}E^*P = \tilde{Z}$$

is c–stable and

(10.16) $$\tilde{Q} + \tilde{A}^*E^{-*}E^*P = -E^*PE^{-1}\tilde{A} + E^*PE^{-1}BR^{-1}B^*E^{-*}E^*P.$$

Setting $\tilde{P} = PE^{-1}$ we get

(10.17) $$\tilde{Q} + \tilde{A}^*\tilde{P}E + E^*\tilde{P}\tilde{A} - E^*\tilde{P}BR^{-1}B^*\tilde{P}E = 0,$$

i.e. $\tilde{P}$ solves (10.4).

From (10.15) we then get that

$$E^{-1}(A - BR^{-1}(B^*\tilde{X}E + S^*C) - BR^{-1}B^*\tilde{P}E) = \tilde{Z}$$

is c–stable and thus $\tilde{P}$ is the correct stabilizing feedback to modify $\tilde{X}$.

We can summarize these observations as follows:

10.18 THEOREM. *Let the columns of* $\begin{bmatrix} Z_1 \\ Z_2 \end{bmatrix} \in \mathbb{C}^{2n,n}$ *span the deflating subspace corresponding to the stable eigenvalues of the pencil*

(10.19) $$\begin{bmatrix} \tilde{A} & BR^{-1}B^* \\ \tilde{Q} & -\tilde{A}^* \end{bmatrix} - \lambda \begin{bmatrix} E & 0 \\ 0 & E^* \end{bmatrix},$$

where $\tilde{A}, \tilde{Q}$ *are as in Theorem 10.3 and let* $\tilde{P} = -Z_2 Z_1^{-1}E^{-1}$, *then* $\tilde{X} + \tilde{P}$ *is the unique Hermitian positive semidefinite solution of (10.1).*

In the discrete case we proceed analogously by considering the deflating subspace of the pencil

(10.20) $$\alpha \begin{bmatrix} A - BR^{-1}S^*C & 0 \\ C^*(Q - SR^{-1}S^*)C & E^* \end{bmatrix} - \beta \begin{bmatrix} E & -BR^{-1}B^* \\ 0 & (A - BR^{-1}S^*C)^* \end{bmatrix}$$

corresponding to the stable eigenvalues (inside the unit circle). If (E, A, B) is strongly stabilizable, then by Corollary 4.39 it follows that this pencil has exactly n stable eigenvalues and by Lemma 10.10 the same holds for the pencil

(10.21) $$\alpha \begin{bmatrix} \hat{A} & 0 \\ \hat{Q} & E^* \end{bmatrix} - \beta \begin{bmatrix} E & -BR^{-1}B^* \\ 0 & \hat{A}^* \end{bmatrix}$$

with $\hat{A}, \hat{Q}$ as in Theorem 10.5.

Let again the columns of $\begin{bmatrix} Z_1 \\ Z_2 \end{bmatrix} \in \mathbb{C}^{2n,n}$ span this stable deflating subspace, i.e.

$$(10.22) \qquad \begin{bmatrix} \hat{A} & 0 \\ \hat{Q} & E^* \end{bmatrix} \begin{bmatrix} Z_1 \\ Z_2 \end{bmatrix} = \begin{bmatrix} E & -BR^{-1}B^* \\ 0 & \hat{A}^* \end{bmatrix} \begin{bmatrix} Z_1 \\ Z_2 \end{bmatrix} Z.$$

By Lemma 10.10 $(E, \hat{A}, B)$ is strongly stabilizable, hence again the standard theory may be applied to $(E^{-1}\hat{A}, E^{-1}B)$, e.g. Sage [S 1]. Let $V = -Z_2 Z_1^{-1}$, then we have that

$$(10.23) \qquad \begin{bmatrix} \hat{A} & 0 \\ \hat{Q} & E^* \end{bmatrix} \begin{bmatrix} I \\ -V \end{bmatrix} = \begin{bmatrix} E & -BR^{-1}B^* \\ 0 & \hat{A}^* \end{bmatrix} \begin{bmatrix} I \\ -V \end{bmatrix} Z$$

or equivalently

$$(10.24) \qquad \begin{bmatrix} E^{-1}\hat{A} & 0 \\ \hat{Q} & I \end{bmatrix} \begin{bmatrix} I \\ -E^*V \end{bmatrix} = \begin{bmatrix} I & -E^{-1}BR^{-1}B^*E^{-*} \\ 0 & -\hat{A}^*E^{-*} \end{bmatrix} \begin{bmatrix} I \\ -E^*V \end{bmatrix} Z.$$

Setting $\hat{V} = VE^{-1}$ we obtain as in Section 6 that $\hat{V}$ solves (10.6) and

$$(10.25) \qquad (I + B\hat{R}^{-1}B^*\hat{V})^{-1}(A - B\hat{R}^{-1}(B^*\hat{X}A + S^*C)) = EZ.$$

Now we employ the following Lemma.

10.26 LEMMA. Let $B, \hat{X}, \hat{R}$ be as in Theorem 10.5, and let $X = \hat{V} + \hat{X}$. Then

$$(10.27) \qquad (I + B\hat{R}^{-1}B^*\hat{V})^{-1}B\hat{R}^{-1} = B(R + B^*XB)^{-1}$$

and

$$(10.28) \qquad (I + B\hat{R}^{-1}B^*\hat{V})^{-1}(I - B\hat{R}^{-1}B^*\hat{X}) = I - B(R + B^*XB)^{-1}B^*X.$$

PROOF: Formula (10.27) is equivalent to

$$B\hat{R}^{-1}(R + B^*XB) = (I + B\hat{R}^{-1}B^*V)B,$$

which holds trivially, since $R + B^*XB = \hat{R} + B^*\hat{V}B$.
For formula (10.28) use again the Sherman–Morrison–Woodbury formula (e.g. Golub/Van Loan [G 12]) to prove that

$$B(R + B^*XB)^{-1}B^*X = BR^{-1}(I + B^*XBR^{-1})^{-1}B^*X = I - (I + BR^{-1}B^*X)^{-1}.$$

Thus, we have to show that

$$(10.29) \qquad (I + B\hat{R}^{-1}B^*\hat{V})^{-1}(I - B\hat{R}^{-1}B^*\hat{X}) = (I + BR^{-1}B^*X)^{-1}$$

or equivalently

$$(10.30) \qquad (I - B\hat{R}^{-1}B^*\hat{X})(I + BR^{-1}B^*X) = I + B\hat{R}^{-1}B^*\hat{V}.$$

Now $I - B\widehat{R}^{-1}B^*\widehat{V} = I - BR^{-1}(I + B^*\widehat{X}BR^{-1})^{-1}\widehat{X} = (I + BR^{-1}B^*\widehat{X})^{-1}$, so (10.30) is equivalent to

(10.31) $$I + BR^{-1}B^*X = (I + BR^{-1}B^*\widehat{X})(I + B\widehat{R}^{-1}B^*\widehat{V})$$

or

(10.32) $$BR^{-1}B^*\widehat{V} = B\widehat{R}^{-1}B^*\widehat{V} + BR^{-1}B^*\widehat{X}BR^{-1}B^*\widehat{V}$$

so (10.31) is equivalent to

(10.33) $$B(R^{-1} - \widehat{R}^{-1} + R^{-1}B^*\widehat{X}B\widehat{R}^{-1})B^*\widehat{V} = 0.$$

But the middle term vanishes again trivially by the Sherman–Morrison–Woodbury formula. ∎

Now Lemma 10.26 implies that (10.25) is equivalent to

$$(I - B(R + B^*XB)^{-1}B^*X)A - B(R + B^*XB)^{-1}S^*C = EZ$$

hence $X = \widehat{V} + \widehat{X}$ is the stabilizing solution of the Riccati equation (10.2).

Based on the results in this Section we have the following general defect correction algorithms, (e.g. Mehrmann/Tan [M 10]). Let $\|\cdot\|$ be a matrix norm that is easily computable and let ϵ be an acceptable, userdefined accurary bound for the residual.

10.34 ALGORITHM CDCORC [COMPLEX DEFECT CORRECTION FOR THE CONTINUOUS TIME ALGEBRAIC RICCATI EQUATION GIVEN BY (10.1)].

Given A, B, C, E, Q, R, S as in (10.1), the following algorithm iteratively corrects the solution of (10.1) by solving the defect equation, and at the same time computes an error estimate $P \in C^{n,n}$ for the solution.

Step 1: Apply any method to compute the "stabilizing" solution $\widetilde{X}$ of (10.1).

Step 2: Set $P := \widetilde{X}, \widetilde{X} := 0$.

Step 3:

    WHILE $\|P\| > \epsilon$

        Set

$$\widetilde{X} := \widetilde{X} + P$$
$$Q := C^*QC + A^*\widetilde{X}E + E^*\widetilde{X}A - (B^*\widetilde{X}E + S^*C)^*R^{-1}(B^*\widetilde{X}E + S^*C)$$
$$A := A - BR^{-1}(B^*\widetilde{X}E + S^*C).$$

        Apply any method to obtain an approximate stabilizing solution of

(10.35) $$Q + A^*PE + E^*PA - E^*PBR^{-1}B^*PE = 0$$

END WHILE

END

It is obvious how to obtain a real version of this method, which we call DCORC. In the discrete case we have the analogous method.

10.36 ALGORITHM CDCORD [COMPLEX DEFECT CORRECTION FOR THE DISCRETE TIME ALGEBRAIC RICCATI EQUATION (10.2)].

Given A, B, C, E, Q, R, S as in (10.2) the following algorithm iteratively corrects the solution of (10.2) by solving the defect equation and at the same time computes an error estimate $V \in \mathbb{C}^{n,n}$ for the solution.

Step 1: Apply any method to compute the "stabilizing" solution $\widehat{X}$ of (10.2).

Step 2: Set $V := \widehat{X}, \widehat{X} := 0$.

Step 3:

While $\|V\| > \epsilon$

Set

$$\widehat{X} := \widehat{X} + V$$
$$Q := -E^*\widehat{X}E + C^*QC + A^*\widehat{X}A-$$
$$(A^*\widehat{X}B + C^*S)(R + B^*\widehat{X}B)^{-1}(A^*XB + C^*S)^*$$
$$A := -B(R + B^*\widehat{X}B)^{-1}S^*C + (I - B(R + B^*\widehat{X}B)^{-1}B^*\widehat{X})A$$
$$R := R + B^*\widehat{X}B$$

Apply **any** method to obtain an approximate stabilizing solution of

$$(10.37) \qquad 0 = Q - E^*VE + A^*VA - (A^*VB)^*(R + B^*VB)^{-1}(A^*VB)$$

END WHILE

END

The corresponding real version of this method we call DCORD. The computational costs for the defect correction is dominated by the cost for the computation of the solution in Step 1 and Step 3. The updating of the matrices can be significantly reduced by observing that a reduction to a Schur–like form allows to read off the residual as well as the updated A. Thus, these updatings essentially do not contribute to the costs if a Schur method is used.

The defect correction algorithms (10.33), (10.35) and their real counterparts are backward numerically stable in the following sense. (See Bunse–Gerstner et al [B 30]). The approximate solution $\widetilde{X}$ to the equation (10.1) is the exact solution of the perturbed equation

$$(10.38) \qquad 0 = C^*QC - \widetilde{Q} + E^*XA + A^*XE - (B^*XE + S^*C)^*R^{-1}(B^*XE + S^*C)$$

where $\widetilde{Q}$ is the residual as in Theorem (10.3). Analogously the approximate solution $\widehat{X}$ to equation (10.2) is the exact solution of the perturbed equation

(10.38)
$$E^*XE - C^*QC - \widehat{Q}$$
$$+ A^*XA - (A^*XB + C^*S)(R + B^*XB)^{-1}(A^*XB + C^*S)^*$$

where $\widehat{Q}$ is the residual as in Theorem 10.5.

If the underlying solution method is sufficiently accurate to get the first significant digits of the defects P or V correct, the approximate solutions $\widetilde{X}, \widehat{X}$ converge to the exact solution. In practice as the iteration converges, the accuracy of the computed residuals $\widetilde{Q}, \widehat{Q}$ declines due to substractive cancellation in the final subtraction of Q since at this point $\widetilde{Q}, \widehat{Q}$ is a rounding error small perturbation of Q. Eventually the errors in the residual affect the most significant digits of the corrections P or V.

Even after this limiting accuray has been reached it is unusual for $\|P\|$ to overestimate $\|X - \widetilde{X}\|$ and for $\|V\|$ to overestimate $\|X - \widehat{X}\|$ by more than a factor of 10. Often it takes only one or two iterations of defect correction to reach limiting accuracy.

Note that there is no requirement that the method used in Step 3 of Algorithms 10.33, 10.35 is the same as that in the corresponding Step 1.

A typical procedure that can be used in Step 3 is Newton's method which is obtained by ignoring the quadratic terms in (10.34), (10.36). Newton's method will be discussed in the next section.

In this section we have shown how to improve solutions of algebraic Riccati equations by iterative refinement. The described procedure can be combined with any solver for these equations and as is suggested in Bunse-Gerstner et al [B 30]: **At least one step of defect correction should be standard procedure for algebraic Riccati equations.**

§ 11 Newton's method.

The algebraic Riccati equation is a quadratic matrix equation, so it is natural to apply Newton's method to obtain an approximate solution.

There exists an extensive literature on the application of Newton's method for the solution of algebraic Riccati equations, for both the continuous and the discrete case.

To name only a few references, e.g. Arnold [A 9], [A 10], Byers [B 39], Hammarling [H 3], Hewer [H 6], Kenney et al [K 16], Kleinman [K 17]. We will review the formulae for the iteration and the methods to compute the iterates. Convergence theorems for both the discrete and continuous case are given in Arnold [A 10], based on the proofs in Kleinman [K 17] and Hewer [H 6]. Unfortunately, though the results are true, the given proofs are partly incorrect. Correct proofs are due to Elsner [E 3] and we give them here. We begin with the continuous problem of computing an Hermitian positive semidefinite solution of (10.1).

We assume that E is nonsingular and obtain the following Newton iteration:

Given $X_0 \in \mathbf{C}^{m,n}$. Define

$$(11.1) \qquad K_j = R^{-1}(B^* X_{j-1} E + S^* C), \quad j = 1, 2, \ldots$$

and X_j as the solution of the Lyapunov equation

$$(11.2) \qquad \begin{aligned} 0 = &(A - BK_j)^* X_j E + E^* X_j (A - BK_j) \\ &+ C^* QC + K_j^* RK_j - C^* SK_j - K_j^* S^* C, \ j = 0, 1, 2, 3, \ldots, \end{aligned}$$

where the numerical solution of the Lyapunov equation (11.2) can be obtained by the method of Hammarling [H 2] or by the Bartels/Stewart algorithm [B 5] as discussed later in this section.

We have the following proof of convergence of Elsner [E 3].

11.3 THEOREM. *Assume that E is nonsingular and $C^*(Q - SR^{-1}S^*)C$ is positive definite. Let X_j, $j = 0, 1, 2, \ldots$ be the unique positive semidefinite solution of (11.2), where X_0 is chosen such that $A - BR^{-1}(B^* XE + S^* C)$ is c-stable, i.e. has all its eigenvalues in the left half plane. Then, we have*

$$(11.4) \qquad 0 \le X \le X_{j+1} \le X_j \le \ldots \le X_0$$

in the canonical order on Hermitian matrices ($X \ge Y$ if $X - Y$ positive semidefinite)

$$(11.5) \qquad A - BR^{-1}(B^* X_j E + S^* C) \quad \text{c-stable}, \ j = 1, 2, \ldots$$

$$(11.6) \qquad \lim_{j \to \infty} X_j = X.$$

Furthermore, there exists a constant γ such that

$$(11.7) \qquad \|X_i - X\| \le \gamma \|X_{i-1} - X\|^2,$$

i.e. we have quadratic convergence.

PROOF: E was assumed nonsingular, so we can define

(11.8)
$$F := E^{-1}(A - BR^{-1}S^*C), \ X := E^*XE,$$
$$H := C^*(Q - SR^{-1}S^*)C, \ G = E^{-1}BR^{-1}B^*E^{-*}.$$

Then (11.2) is equivalent to

(11.9)
$$X_j(F - GX_{j-1}) + (F - GX_{j-1})^*X_j = -H - X_{j-1}GX_{j-1}.$$

Furthermore, we have

(11.10)
$$(F - GX_j)^*X_j + X_j(F - GX_j) =$$
$$- H - X_jGX_j - (X_j - X_{j-1})G(X_j - X_{j-1}), \ j = 1, 2, \ldots$$

and by combining (11.9), (11.10)

(11.11)
$$(F - GX_j)^*(X_j - X_{j+1}) + (X_j - X_{j+1})(F - GX_j) =$$
$$- (X_{j-1} - X_j)G(X_{j-1} - X_j), \ j = 1, 2, \ldots .$$

By applying Lyapunov's theorem, e.g. Gantmacher [G 4] or Horn/Johnson [H 8] we obtain from (11.9) inductively that X_j is positive semidefinite, from (11.10) that $F - GX_j$ is c-stable and it follows from (11.11) that $X_j - X_{j+1}$ is positive definite.

Thus, the X_j form a nonincreasing sequence, which is bounded from below, and is hence convergent by the Bolzano–Weierstraß theorem and furthermore $X = \lim_{j \to \infty} X_j$ is positive semidefinite. By (11.10) it follows that

(11.12)
$$(F - GX)^*X + X(F - GX) = -H - XGX.$$

Thus, X solves (11.1) and is positive semidefinite. Now from (11.12) we have

(11.13) $\quad (F - GX_j)^*X + X(F - GX_j) = -H - XGX + (X - X_j)GX + XG(X - X_j).$

Subtracting (11.10) from (11.13) we obtain

$$(F - GX_j)^*(X - X_j) + (X - X_j)(F - GX_j) =$$
$$- XGX + (X - X_j)GX + XG(X - X_j) + X_jGX_j$$
$$+ (X_j - X_{j-1})G(X_j - X_{j-1}) = (X - X_j)G(X - X_j)$$
$$+ (X_j - X_{j-1})G(X_j - X_{j-1})$$

and therefore

(11.14)
$$(F - GX)^*(X - X_j) + (X - X_j)(F - GX) =$$
$$- (X - X_j)G(X - X_j) + (X_j - X_{j-1})G(X_j - X_{j-1}).$$

This implies

$$0 \leq X_j - X \leq \int_0^\infty e^{t(F-GX)^*}[(X_j - X_{j-1})G(X_j - X_{j-1})]e^{t(F-GX)^*}dt$$

(11.15)
$$\leq \|X_j - X_{j-1}\|^2 \int_0^\infty e^{t(F-GX)}Ge^{t(F-GX)}dt$$

$$=: \|X_j - X_{j-1}\|^2 Y.$$

Thus, we obtain

(11.16) $\|X_j - X\| \leq \|Y\| \, \|X_j - X_{j-1}\|^2 = \gamma\|X_j - X_{j-1}\|^2 \leq \gamma\|X - X_{j-1}\|^2,$

since $X_{j-1} - X \geq X_{j-1} - X_j \geq 0$. ∎

In the discrete case we proceed in an analogous way, assuming again that E is nonsingular. Then we have the following Newton iteration for (10.2):

Given $X_0 \in \mathbb{C}^{m,n}$, define

(11.17) $K_j = (R + B^*X_{j-1}B)^{-1}(B^*X_{j-1}A + S^*C), \; j = 1, 2, \ldots$

and compute X_j as the solution of the Stein equation

(11.18)
$$E^*X_jE = (A - BK_j)^*X_j(A - BK_j) + K_j^*RK_j$$
$$+ C^*QC - C^*SK_j - K_j^*S^*C, \; j = 0, 1, 2, \ldots.$$

We have the following proof of convergence, e.g. Elsner [E 3]:

11.19 THEOREM. *Assume that E is nonsingular and $C^*(Q - SR^{-1}S^*)C$ positive definite. Let X_j be the unique positive semidefinite solution of (11.2) where X_0 is chosen such that $Z_0 = E^{-1}(A - B(R + B^*X_0B)^{-1}(B^*X_0A + S^*C))$ is d–stable, i.e. has all eigenvalues inside the unit circle or in other words $\rho(Z_0) < 1$, where $\rho(Z_0)$ denotes the spectral radius of Z_0. Then*

(11.20) $0 \leq X \leq X_{j-1} \leq X_j \leq \ldots \leq X_0, \quad j = 0, 1, \ldots$

(11.21) $Z_j = E^{-1}(A - B(R + B^*X_jB)^{-1}(B^*X_jA + S^*C))$ *is d–stable*

(11.22) $\lim_{j \to \infty} X_j = X.$

Furthermore, there exists a constant γ such that

(11.23) $\|X_{j+1} - X\| \leq \gamma\|X_j - X\|^2$

i.e. we have quadratic convergence.

PROOF: Setting

(11.24) $$X_j := E^* X_j E, \ A := E^{-1}A, \ B := E^{-1}B,$$

it follows that (11.18) is equivalent to

(11.25)
$$\begin{aligned}
&X_j - (A - BK_j)^* X_j (A - BK_j) + K_j^* RK_j \\
&+ C^* QC - C^* SK_j - K_j^* S^* C = C^* (Q - SR^{-1}S^*)C \\
&+ (K_j^* R - C^* S)R^{-1}(K_j^* R - C^* S)^*.
\end{aligned}$$

Now let $\widehat{K}_1, \widehat{K}_2$ and $\widehat{X}_1, \widehat{X}_2$ satisfy

$$\widehat{X}_i = (A - B\widehat{K}_i)^* \widehat{X}_i (A - B\widehat{K}_i) + \widehat{K}_i^* R\widehat{K}_i + C^* QC - C^* S\widehat{K}_j - \widehat{K}_j^* S^* C, \ i = 1, 2, \dots .$$

Then

(11.26)
$$\begin{aligned}
\widehat{X}_1 - \widehat{X}_2 &= (A - B\widehat{K}_2)^* (\widehat{X}_1 - \widehat{X}_2)(A - B\widehat{K}_2) \\
&+ (\widehat{K}_1 - \widehat{K}_2)^* (R + B^* \widehat{X}_2 B)(\widehat{K}_1 - \widehat{K}_2) \\
&+ [-C^* S + \widehat{K}_2 (R + B^* \widehat{X}_1 B) - B^* \widehat{X}_1 A](\widehat{K}_1 - \widehat{K}_2) \\
&+ (\widehat{K}_1 - \widehat{K}_2)^* [(R + B^* \widehat{X}_1 B)\widehat{K}_2 - A^* \widehat{X}_1 B - S^* C].
\end{aligned}$$

Apply (11.26) with $\widehat{X}_1 = X_0$, $\widehat{X}_2 = X$, $\widehat{K}_1 = K_0$, $\widehat{K}_2 = K$. Then

(11.27) $\quad X_0 - X = (A - BK_0)^*(X_0 - X)(A - BK_0) + (K_0 - K)^*(R + BXB)(K_0 - K),$

where other terms vanish by (11.17).

Applying the discrete version of Lyapunov's Theorem, e.g. Lancaster/Tismenetsky [L 5], it follows that $X_0 - X$ is positive semidefinite, since $A - BK_0$ is d–stable.

Now

(11.28)
$$\begin{aligned}
&X_j - (A - BK_{j+1})^* X_j (A - BK_{j+1}) - K_{j+1}^* RK_{j+1} \\
&- C^* QC + C^* SK_{j+1} + K_{j+1}^* S^* C = X_j - X_{j+1} \\
&- (A - BK_{j+1})^* (X_j - X_{j+1})(A - BK_{j+1})
\end{aligned}$$

by (11.18) for $j + 1$.

Applying again (11.26) with $\widehat{X}_1 = X_j$, $\widehat{X}_2 = X_{j+1}$, $\widehat{K}_2 = K_{j+1}$, $\widehat{K}_1 = K_j$ and (11.17), we obtain for $j = 0, 1, 2, \dots$ from (11.25) that

(11.29)
$$\begin{aligned}
&X_j - (A - BK_{j+1})^* X_j (A - BK_{j+1}) \\
&= C^* QC - C^* SK_{j+1} - K_{j+1}^* S^* C \\
&+ K_{j+1}^* RK_{j+1} + (K_j - K_{j+1})^* (R + B^* X_j B)(K_j - K_{j+1}) \\
&= (K_j - K_{j+1})^* (R + B^* X_j B)(K_j - K_{j+1}) \\
&+ C^* (Q - SR^{-1}S^*)C + (K_{j+1}^* R - C^* S)R^{-1}(K_{j+1}^* R - C^* S)^*
\end{aligned}$$

and the right side is positive definite by assumption. Thus, again by the discrete Lyapunov Theorem it follows inductively that $Z_j = A - BK_j$ d–stable implies X_j positive definite, which then by (11.26) implies that $Z_{j+1} = A - BK_{j+1}$ is d–stable and then by (11.25) again X_{j+1} positive definite. By again applying (11.26) we obtain (11.21). Then, since $\lim\limits_{j\to\infty} X_j = X_\infty$ exists we have $X_\infty \geq X$. It follows from (11.26) that $X - X_\infty \geq 0$. Thus, $X = X_\infty$.

To prove (11.23), we consider

$$0 \leq X_i - X = \sum_{j=0}^{\infty}((A - BK_i)^*)^j(K_i - K)^*(R + B^*XB)(K_i - K)(A - BK_i)^j$$

$$\leq \|K_i - K\|^2\|R + B^*XB\| \sum_{j=0}^{\infty}((A - BK_i)^*)^j(A - BK_i)^j.$$

Now $A - BK_j$ is d–stable and thus $A - BK$ is d–stable. Thus, the sequence of operators L_i^{-1} defined by $L_iX = X - (A - BK_i)^*X(A - BK_i)$ and its limit L^{-1} are uniformly bounded by a constant C. Hence by (11.26)

$$\|X_i - X\| \leq C\|K_i - K\|^2 \leq \gamma\|X_{i-1} - X\|^2.$$

∎

Observe that in both the discrete and continuous case, Newton's method is an approximate defect correction method. If we leave out the quadratic term in P in (10.4) or in V in (10.6), then we obtain directly the Newton iterations (11.2), (11.18), respectively. This treatment is certainly justified if P or V are small, i.e. if we have a good approximation to begin with.

In general Newton's method or variatons of it like the secant method of Dieci et al [D 10] for the solution of algebraic Riccati equations need good starting matrices K_0 to be successful. A method to obtain such starting matrices is for example given in Hammarling [H 3]. In view of Section 10, however, we can see that Newton's method is a very effective method for iterative refinement following another method that produces a good initial K_0, X_0.

Regions of convergence for Newton's method are discussed in Kenney et al [K 16].

For practical implementations of Newtons method in the case $E = I$, see Arnold [A 9], [A 10], Byers [B 39], Hammarling [H 3]. These methods are based on the well–established methods to solve the Lyapunov equation in (11.2) or the Stein equation in (11.18). Let us first discuss the Lyapunov equation in (11.2). It is of the form

(11.30) $$P^*XE + E^*XP = W.$$

If $E = I$, then performing a QR-decomposition $P^* = QR$, we have to solve the equation $R^*\tilde{X} + \tilde{X}R = Q^*WQ$, which can be done by backward substitution as described in the Bartels/Stewart algorithm [B 5]. If $E \neq I$ then we may apply the QZ-algorithm of Moler and Stewart, e.g. Golub/Van Loan [G 12], to $\alpha E - \beta P$ and we obtain Q, Z such that

$Q(\alpha E - \beta P)Z = \alpha T_1 - \beta T_2$ is in generalized Schur–form (real generalized Schur–form). Setting $\widetilde{X} = QXQ^*$, we obtain the new equation

(11.31) $$T_2^* \widetilde{X} T_1 + T_1^* \widetilde{X} T_2 = Z^* W Z = V.$$

If $\alpha E - \beta P$ is c-stable, i.e. has only eigenvalues with negative real part, then T_1 is invertible and the Bartels/Stewart algorithm [B 5] can be applied to

(11.32) $$T_1^{-*} T_2^* \widetilde{X} + \widetilde{X} T_2 T_1^{-1} = T_1^{-*} Z^* W Z T_1^{-1}.$$

Observe that the inversion of T is not really necessary, since it can be done implicitly in the backward substitution process given in the following algorithm.

11.33 ALGORITHM CBSLYA [BARTELS/STEWART ALGORITHM FOR THE SOLUTION OF THE LYAPUNOV EQUATION (11.32)].

Given $T_1 = [t_{ij}]$ and $T_2 = [s_{ij}]$ upper triangular and $V = Z^* W Z = [v_{ij}]$, $T_1, T_2, V \in \mathbb{C}^{n,n}$, the following algorithm overwrites V with the solution X of (11.32).

FOR $k = 1, \dots, n$
    FOR $j = 1, \dots, n$
        FOR $i = 1, \dots, j$
            $v_{kj} = v_{kj} - s_{kk} v_{ki} t_{ij} + t_{jj} v_{ki} s_{ij}$
        END FOR
        $v_{kj} = v_{kj}/(s_{kk} t_{jj} + s_{jj} t_{kk})$
    END FOR
    END FOR
END

We then have the following Newton iteration

11.34 ALGORITHM CNEWTONC [NEWTONS METHOD FOR THE SOLUTION OF THE CONTINUOUS TIME ALGEBRAIC RICCATI EQUATION].

Given matrices A, B, C, E, Q, R, S as in (1.1) and a starting matrix X_0 such that $K_0 = R^{-1}(B^* X_0 E + S^* C)$ is c-stable, this algorithm performs the Newton iteration (11.2), (11.3).

FOR $j = 1, 2, \dots$, UNTIL SATISFIED

    Set $K_j := R^{-1}(B^* X_{j-1} E + S^* C)$, $P := A - BK_j$, $W = C^* QC + K_j^* RK_j - C^* SK_j - K_j^* S^* C$.

    Apply the QZ–algorithm to $\alpha E - \beta P$ and compute unitary Q, Z such that $Q(\alpha E - \beta P)Z = \alpha T_1 - \beta T_2$ is in generalized Schur–form.

    Solve equation (11.31) by Algorithm 11.33 CBSLYA applied to $T_1, T_2, V = Z^* W Z$.

    Set $X_j = Q^* \widetilde{X} Q$.

END FOR

END

In the real case we call this method NEWTONC and use a real version of the Bartels/Stewart method based on quasitriangular matrices called BSLYA.

For the discrete time problem we have to solve the Stein equation

$$(11.35) \qquad E^*XE = P^*XP + W.$$

Using the QR-algorithm of Francis [F 4], [F 5] or the QZ-algorithm of Moler and Stewart, e.g. Golub/Van Loan [G 12], we perform a Schur decomposition or generalized Schur decomposition of $\alpha E - \beta P$ depending on whether $E = I$ or not, such that

$$(11.36) \qquad \alpha E - \beta P = \alpha Q^* T_2 Z - \beta Q^* T_1 Z,$$

where T_1, T_2 are upper triangular.
Then we have the new equation for $\tilde{X} = QXQ^*$, $V = ZWZ^*$,

$$(11.37) \qquad T_2^* \tilde{X} T_2 = T_1^* \tilde{X} T_1 + V.$$

We then have the following variation of the Bartels/Stewart algorithm.

11.38 ALGORITHM CBSSTE [BARTELS/STEWART ALGORITHM FOR THE SOLUTION OF THE STEIN EQUATION (11.37)].

Given $T_1 = [t_{ij}]$, $T_2 = [s_{ij}]$ upper triangular and $V = [v_{ij}]$, $T_1, T_2, V \in \mathbb{C}^{n,n}$, the following algorithm overwrites V with the solution $\tilde{X}$ of (11.37).

    FOR $k = 1, \ldots, n$
        FOR $j = 1, \ldots, n$ b
            FOR $i = 1, \ldots, j$
                $v_{kj} = v_{kj} - s_{kk} v_{ki} s_{ij} - t_{kk} v_{ki} t_{ij}$
            END FOR
            $v_{kj} = v_{kj}/(s_{kk} s_{jj} + t_{kk} t_{jj})$
        END FOR
    END FOR
END

Newton's method in the discrete case is then as follows:

11.39 ALGORITHM CNEWTOND [NEWTONS METHOD FOR THE SOLUTION OF THE DISCRETE TIME ALGEBRAIC RICCATI EQUATION].

Given matrices A, B, C, E, Q, R, S as in (1.2) and a starting matrix X_0 such that $K_0 = (R + B^* X_0 B)^{-1}(B^* X_0 E + S^* C)$ is d-stable, this algorithm performs the Newton iteration (11.17), (11.18).

    FOR $j = 1, 2, \ldots$, UNTIL SATISFIED

Set $K_j := (R + B^* X_{j-1} B)^{-1} (B^* X_{j-1} E + S^* C)$, $P := A - BK_j$, $W = C^* QC + K_j^* RK_j - C^* SK_j - K_j^* S^* C$.

Apply the QZ-algorithm to $\alpha E - \beta \Gamma$ and compute unitary Q, Z such that $Q(\alpha E - \beta P)Z = \alpha T_1 - \beta T_2$ is in generalized Schur–form.

Solve equation (11.37) by Algorithm 11.38 CBSSTE applied to $T_1, T_2, V = Z^* W Z$.

Set $X_j = Q^* \tilde{X} Q$.

    END FOR

END

Again we immediately obtain a real version of this method, NEWTOND, using a real version BSSTE of Algortihm 11.38 based on quasitringular matrices.

We see that under the assumption of a c–stable (d–stable) P both algorithms do not fail.

The computational cost per step of Newton's method is in both the continuous and discrete case dominated by the computation of the generalized Schur form or the computation of the QR-decomposition in the continuous case with $E = I$. These are all algorithms that need $0(n^3)$ flops, for the exact figures see Golub/Van Loan [G 12].

Only with a very good starting guess, is Newton's method competitive with the other methods as a direct method.

In this section we have described Newton's method for the solution of the discrete and continuous algebraic Riccati equation and given proofs of convergence, provided good starting matrices are given. If this is the case, and we might obtain such an initial matrix by any of the other methods for the solution of algebraic Riccati equations, then Newton's method is a very good procedure for the iterative refinement of the solution.

§ 12 The sign function method.

The matrix sign function method is an elegant and, when combined with defect correction, effective numerical method for the algebraic Riccati equation. It has been studied for example in Balzer [B 1], Barraud [B 3], [B 4], Beavers/Denman [B 6], [B 7], [B 8], [B 9], Byers [B 39], [B 44], Denman/Beavers [D 7], Gardiner/Laub [G 6], Howland [H 10], Kenney/Laub [K 11], [K 12], [K 13], [K 14], Roberts [R 9].

A recent survey about the matrix sign function method for solving Riccati equations is given in Bunse–Gerstner et al [B 30].

Let $A \in \mathbf{C}^{n,n}$ and let $P \in \mathbf{C}^{n,n}$ be nonsingular such that $J = PAP^{-1}$ is in Jordan canonical form.

Let $J = D + N$, where $D = \mathrm{diag}(d_1, \dots, d_n)$ is the diagonal of J and N the rest and let

$$(12.1) \qquad \qquad \mathrm{sign}(A) = P^{-1} \Sigma P,$$

where $\Sigma = \mathrm{diag}(\sigma_1, \dots, \sigma_n)$ and

$$(12.2) \qquad \qquad \sigma_i = \left\{ \begin{array}{l} +1 \text{ if } Re(d_i) > 0 \\ -1 \text{ if } Re(d_i) < 0 \end{array} \right\} \quad i = 1, \dots, n.$$

If A has purely imaginary eigenvalues, then $\mathrm{sign}(A)$ is not defined. It follows immediately from the definition that if $\mathrm{sign}(A)$ is defined and P is nonsingular, then

$$\mathrm{sign}(P^{-1}AP) = P^{-1} \mathrm{sign}(A)P.$$

Clearly, $\mathrm{sign}(A)$ is a square root of I. Moreover, if A has no eigenvalue on the imaginary axis, then the Newton iteration for $\sqrt{I}$,

$$(12.3) \qquad \qquad Z_0 := A$$

$$(12.4) \qquad \qquad Z_{k+1} = Z_k - \frac{Z_k - Z_k^{-1}}{2} = \frac{Z_k + Z_k^{-1}}{2}$$

satisfies

$$(12.5) \qquad \qquad \lim_{k \to \infty} Z_k = \mathrm{sign}(A).$$

Now consider the algebraic Riccati equation

$$(12.6) \qquad \begin{aligned} 0 &= E^*XA + A^*XE + C^*QC - (B^*XE + S^*C)^*R^{-1}(B^*XE + S^*C) \\ &= C^*(Q - SR^{-1}S^*)C + E^*X(A - BR^{-1}S^*C) + (A - BR^{-1}S^*C)^*XE \\ &\quad - E^*XBR^{-1}B^*XE. \end{aligned}$$

If E is nonsingular, as we may assume by Section 9 we can make the replacements

$$(12.7) \qquad \begin{aligned} & X := E^*XE, \ F := E^{-1}(A - BR^{-1}S^*C) \\ & G := E^{-1}BR^{-1}B^*E^{-*}, H := C^*(Q - SR^{-1}S^*)C. \end{aligned}$$

Then (12.6) is equivalent to the equation:

$$(12.8) \qquad K = \begin{bmatrix} F & G \\ H & -F'^* \end{bmatrix} = \begin{bmatrix} I & 0 \\ -X & I \end{bmatrix} \begin{bmatrix} F - GX & G \\ 0 & -(F - GX)^* \end{bmatrix} \begin{bmatrix} I & 0 \\ -X & I \end{bmatrix}^{-1} .$$

If X is the stabilizing solution of (12.6), i.e. $F - GX$ c–stable, then the sign function is defined for K and applying it, we obtain

$$(12.9) \qquad W = \begin{bmatrix} W_{11} & W_{12} \\ W_{21} & W_{22} \end{bmatrix} := \text{sign}(K) = \begin{bmatrix} I & 0 \\ -X & I \end{bmatrix} \begin{bmatrix} -I & Z \\ 0 & I \end{bmatrix} \begin{bmatrix} I & 0 \\ -X & I \end{bmatrix}^{-1} ,$$

where Z satisfies the Lyapunov equation

$$(12.10) \qquad (F - GX)Z + Z(F - GX)^* = -2H.$$

Thus, X is the solution of the overdetermined consistent system of linear equations

$$(12.11) \qquad \begin{bmatrix} W_{12} \\ W_{22} + I \end{bmatrix} X = \begin{bmatrix} W_{11} + I \\ W_{21} \end{bmatrix} .$$

Using the QR-factorization of $\begin{bmatrix} W_{12} \\ W_{22} + I \end{bmatrix}$, e.g. Golub/Van Loan [G 12] we can compute X.

For the continuous Riccati equation, the iteration (12.3), (12.4) is applied to a Hamiltonian starting value $Z_0 = \mathcal{H}$. It follows that each iterate Z_k is Hamiltonian. Replacing the non–symmetric matrix inversion Z_k^{-1} by the symmetric matrix inversion $(JZ_k)^{-1} J$, cuts the work and storage costs in half, see Byers [B 43]. Scaling strategies to accelerate convergence have been studied in Balzer [B 1], Barraud [B 3], Byers [B 43]. Perhaps the best strategy suited to Riccati equations is that of Byers [B 43]. There, it is suggested that Z_k be scaled to have determinant 1 at each iteration. A fortuitous side effect of this scaling strategy is that in the scalar case ($n = 1$) the iteration degenerates to the ordinary scalar quadratic formula. In the $n = 2$ case, if the eigenvalus of the initial Hamiltonian matrix H are real, then the iteration converges to $\text{sign}(H)$ in exactly 2 iterations, thus giving a generalization of the quadratic formula to this special case.

The following algorithm solves the continuous algebraic Riccati equation

$$(12.12) \qquad 0 = XF + F^*X + H - XGX$$

with the matrix sign function method.

12.13 ALGORITHM CSIGNF [COMPLEX SIGN FUNCTION METHOD].

Given F, H, G as in (12.7) such that $\begin{bmatrix} F & G \\ H & -F^* \end{bmatrix}$ has no purely imaginary eigenvalues, the following algorithm computes the solution of (12.12) and an error estimate $P \in C^{n,n}$.

Step 1 $\quad Z_0 = \begin{bmatrix} F & G \\ H & -F^* \end{bmatrix} .$

Step 2

FOR $k = 0, 1, 2, \ldots$ UNTIL Z_k converges

Set $Z_k = |\det Z_k|^{-\frac{1}{2n}} Z_k$.

Set $Z_{k+1} = Z_k - \frac{1}{2}(Z_k - (JZ_k)^{-1})J)$.

END FOR

Step 3 Solve the linear system (12.11) for X.

Step 4

Use the defect correction algorithm 10.34 CDCORD to refine the solution and to produce an error estimate P.

END

The real version of this method is called SIGNF.

The determinant in Step 2 is a by product of the factorization used to invert JZ_k in Step 2, so it is essentially free. See, for example, the discussion of DSICO and DSIDI in Dongarra et al [D 11]. The consistent, overdetermined system of equations is better solved with a QR-factorization than through the normal equations, e.g. Lawson et al [L 15]. See also the discussion of subroutines DQRDC and DQRSL, Dongarra et al [D 11]. In practical experience, the iteration converges to working precision in eight or nine steps and the computational cost per step is $\mathcal{O}(4n^3)$. An extensive study of the convergence behaviour of (12.4) appears in Barraud [B 3].

Barraud [B 3] and Gardiner and Laub [G 6] extended the matrix sign function to matrix pencils, so that it can be applied to the discrete algebraic Riccati equation, too. The main idea in the discrete case is to use the Cayley transformation (Theorem 7.12) to transform the problem to a Hamiltonian problem. From the numerical point of view this approach seems (at least for the time being) not advisable, since the use of the Cayley transformation introduces extra roundoff before starting the process. Also it is quite expensive to perform this transformation.

It is difficult to asses the affect of rounding errors in Algorithm 12.13. Some progress has been made by Byers [B 44], but the issue is still not clear. Thus, to insure numerical stability, Algorithm 12.13 should contain at least one step of defect correction.

In Gardiner/Laub [G 6] it has been shown how the sign function method can be used on modern supercomputers. At this time the sign function method is essentially the only method that has this potential, i.e. can be applied to large problems.

The sign function method is an effective and when combined with defect correction, reliable numerical method to compute the solution of algebraic Riccati equations in the continuous time case. In the discrete time case, there is research in progress to achieve similar properties.

§ 13 Elementary transformation matrices.

In the following sections we discuss factorization methods that transform matrices or pencils to the Hessenberg and Schur forms discussed in Section 7.

These transformations are obtained using elementary transformations like Householder reflections, Givens rotations, e.g. Householder [H 9], Golub/Van Loan [G 12], Householder symplectic matrices, Givens symplectic matrices e.g. Paige/Van Loan [P 2] or elementary symplectic transformations, e.g. Bunse–Gerstner/Mehrmann [B 32], Bunse–Gerstner et al [B 36]. In this Section we give a survey of these simple transformations and describe algorithms that generate them.

1) Transformations in $\mathcal{U}_n(\mathbf{C})(\mathcal{U}_n(\mathbf{R}))$. (E.g. Golub/Van Loan [G 12])

A *Householder* matrix has the form $I - 2\frac{ww^*}{w^*w}$, where $w \in \mathbf{C}^n$, $(w \in \mathbf{R}^n)$.

13.1 ALGORITHM CHHGEN [GENERATION OF COMPLEX HOUSEHOLDER ELIMINATION MATRICES].

Given $x \in \mathbf{C}^n$ and integers k, j, $1 \leq k \leq j \leq n$ the following algorithm determines a Householder vector $w \in \mathbf{C}^n$ or in other words the corresponding matrix

$$(13.2) \qquad H(k,j,w) = I - \beta ww^* = CHHGEN(k,j,x),$$

such that

$$\beta = \frac{2}{w^*w}, \quad w^T = [0 \quad \dots \quad 0 \quad w_k \quad \dots \quad w_j \quad 0 \quad \dots \quad 0]$$

and the components $k+1$ through j of $H(k,j,w)x$ are zero.

$\quad w = 0$

$\quad \mu = \max \{|x_k|, \dots, |x_j|\}$

$\quad \alpha = 0$

$\quad \text{FOR} \quad i = k, j$

$\qquad w_i = x_i / \mu$

$\qquad \alpha = \alpha + |w_i|^2$

$\quad \text{END FOR}$

$\quad \alpha = \sqrt{\alpha}$

$\quad \beta = 1/(\alpha(\alpha + |w_k|))$

$\quad w_k = w_k + \alpha \frac{w_k}{|w_k|}.$

END

In the real case we have the analogous algorithm called HHGEN. (E.g. Golub/Van Loan [G 12]).

A *Givens* matrix has the form

$$(13.3) \qquad G(k,j,c,s) = I + (\bar{c} - 1)e_k e_k^* + (c - 1)e_j e_j^* + \bar{s}e_k e_j^* - se_j e_k^*,$$

where $c, s \in \mathbf{C}$, $|c|^2 + |s|^2 = 1$.

13.4 Algorithm CGIVGEN [Generation of Givens elimination matrix].

Given $x \in \mathbf{C}^n$ and integers k, j satisfying $1 \leq k \leq j \leq n$ the following algorithm determines c, s or in other words the corresponding matrix

(13.5) $G(k, j, c, s) = CGIVGEN\ (k, j, x)$

such that $G(k, j, c, s)^* x$ has a zero in component j.

> IF $x_j = 0$ THEN
>
> > $c = 1, s = 0$
>
> ELSE
>
> > IF $|x_j| \geq |x_k|$ THEN
> >
> > > $t = x_k / x_j$, $s = 1/(1 + |t|)^2)^{\frac{1}{2}}$, $c = st$
> >
> > ELSE
> >
> > > $t = x_j / x_k$, $c = 1/(1 + |t|^2)^{\frac{1}{2}}$, $s = ct$
> >
> > END IF
>
> END IF

END

In the real case we have the analogous algorithm called GIVGEN (e.g Golub/Van Loan [G 12]).

2) Transformations in $\mathcal{US}_{2n}(\mathbf{C})$, $\mathcal{US}_{2n}(\mathbf{R})$. (E.g. Paige/Van Loan [P 2]).

A *Householder symplectic* matrix has the form

(13.6) $\begin{bmatrix} H(k, j, w) & 0 \\ 0 & H(k, j, w) \end{bmatrix}$

with $H(k, j, w)$ defined as in (13.2).

The generation of such a matrix is performed via Algorithm 13.1.

A *Givens symplectic* matrix has the form

(13.7) $G(k, j, c, s) = I + (c - 1)(e_j e_j^T + e_k e_k^T) + s(e_j e_k^T - e_k e_j^T),$

where furthermore $G(k, j, c, s) \in \mathbf{C}^{2n, 2n}$, $j = n + k$, $\bar{s}c \in \mathbf{R}$.

c, s are generated by Algorithm 13.3 but CGSGEN $(k, x) = G(k, k + n, c, s)$ has the particular property, which the initial data have to satisfy, that $\bar{s}c \in \mathbf{R}$.

Note that Givens symplectic matrices are often called Jacobi symplectic matrices (Paige/Van Loan [P 2]).

In the real case the procedure is called GSGEN.

3) Nonorthogonal transformations in $S_{2n}(\mathbf{R})$ (E.g. Bunse–Gerstner/Mehrmann [B 32] and Bunse–Gerstner et al [B 36]).

For $k \in \mathbf{N}$, $1 \leq k \leq n$, $v \subset \mathbf{R}$ define

$$(13.8) \qquad G_0(k, d, v) = \begin{bmatrix} D & DV_0 \\ 0 & D^{-1} \end{bmatrix} = \begin{bmatrix} D & 0 \\ 0 & D^{-1} \end{bmatrix} \begin{bmatrix} I & V_0 \\ 0 & I \end{bmatrix},$$

where $V_0 = v(e_{k-1}e_k^T + e_k e_{k-1}^T) \in \mathbf{R}^{n,n}$,

$$D = I + (d-1)e_k e_k^T, \quad d = \frac{1}{\sqrt[4]{1+v^2}}.$$

Note that

$$(13.9) \qquad G_0(k, d, v)^{-1} = \begin{bmatrix} D^{-1} & -V_0 D \\ 0 & D \end{bmatrix},$$

and that $G_0(k_o, d, v)$ is optimally conditioned in the following sense:

$$(13.10) \qquad \text{cond}_2(G_0(k, d, v)) = \min_{\substack{D \text{ diagonal} \\ D \in S_{2n}(\mathbf{R})}} \{ \text{cond}_2(DG_0(k, 1, v)) \}.$$

(See Bunse–Gerstner/Mehrmann [B 32]).

13.11 ALGORITHM G0GEND [GENERATION OF OPTIMALLY CONDITIONED ELIMINATION MATRIX OF TYPE G_0].

Given $k \in \mathbf{N}$, $1 < k \leq n$ and $x \in \mathbf{R}^{2n}$, where $x_{n+k-1} = 0$ only if $x_k = 0$, the following algorithm determines

$$(13.12) \qquad G_0(k, d, v) = G0GEND(k, x),$$

such that the k-th component of $G_0(k, d, v)x$ is zero.

> IF $\quad x_k = 0 \quad$ THEN
>
> $\qquad v = 0$
>
> ELSE
>
> $\qquad v = -x_k / x_{k+n-1}$
>
> END IF
>
> $d = 1/\sqrt[4]{1+v^2}$.

END

An overview over the elementary routines is given in the following table, also containing the approximate flop counts for the given routines. In the real case we count real flops and in the complex case, complex flops. (Following C.B. Moler, we define a *flop* to be the computational work to evaluate the FORTRAN statement $S = S + A[I, K] * B[K, J]$.)

13.23 TABLE (ELEMENTARY SUBROUTINES).

name	function genera-tion of	input data	output data	cost of com-putation
CHHGEN	complex House-holder matrices	$x \in \mathbb{C}^n$, $k, j \in \mathbb{N}$	$\beta \in \mathbb{R}$, $w \in \mathbb{C}^n$	$2(j-k)$ flops 1 square root
HHGEN	real Householder matrices	$x \in \mathbb{R}^n$, $k, j \in \mathbb{N}$	$\beta \in \mathbb{R}$, $w \in \mathbb{R}^n$	$2(j-k)$ flops 1 square root
CGIVGEN	complex Givens rotation	$x \in \mathbb{C}^n$, $k, j \in \mathbb{N}$	$c, s \in \mathbb{C}$	4 flops 1 square root
GIVGEN	real Givens rota-tion	$x \in \mathbb{R}^n$, $k, j \in \mathbb{N}$	$c, s \in \mathbb{R}$	4 flops 1 square root
CGSGEN	complex symplec-tic Given rotation	$x \in \mathbb{C}^{2n}$, $k \in \mathbb{N}, \; k \leq n$	$c, s \in \mathbb{C}$ $c\overline{s} \in \mathbb{R}$	4 flops 1 square root
GSGEN	real symplectic Givens rotation	$x \in \mathbb{R}^{2n}$, $k \in \mathbb{N}, \; k \leq n$	$c, s \in \mathbb{R}$	4 flops 1 square root
G0GEND	optimally condi-tioned elimination matrix of type G_0	$x \in \mathbb{R}^{2n}, k \in \mathbb{N}$ $1 \leq k \leq n$ $x_{n+k-1} = 0$ only if $x_k = 0$	$v, d \in \mathbb{R}$	3 flops 1 fourth root

Using these elementary transformations we can now describe the various Schur methods that can be used to compute solutions to algebraic Riccati equation. This is done in the following sections.

§ 14 Schur methods.

In Section 7 we have described several Schur forms for Hamiltonian matrices or symplectic pencils. Numerical methods that achieve these forms are obviously methods to compute the stable invariant subspace of a Hamiltonian matrix or the stable deflating subspace of symplectic pencils, since from the triangular forms it is clear which columns of the transformation marices span the required subspace. For the Schur form of an arbitrary matrix or the generalized Schur form of an arbitrary regular pencil there are well–known and widely used methods available, the QR-algorithm of Francis [F 4], [F 5], e.g. Garbow et al [G 2], Smith et al [S 11], Wilkinson [W 8], Wilkinson/Reinsch [W 10] or Golub/Van Loan [G 12], and the QZ-algorithm of Moler/Stewart [M 13] and Ward [W 2], e.g. Garbow et al [G 2], Smith et al [S 11] or Golub/Van Loan [G 12]. In the following we denote the standard QR-algorithm for real or complex matrices by QR, CQR respective and analogously we denote the QZ-algorithm by QZ, CQZ. If we want to use these methods for the computation of the required stable deflating subspaces for Hamiltonian matrices or symplectic pencils, they have to be combined with an ordering procedure for the eigenvalues on the diagonal, since in order to read off the stable invariant or deflating subspace from the transformation matrix, we need the stable eigenvalues in the upper left part of the Schur form.

This ordering is possible by employing algorithms of Ruhe [R 12], Stewart [S 17] or Bartels/Stewart [B 5], e.g. Golub/Van Loan [G 12]. The application and modification of these two methods for Hamiltonian and symplectic problems was done in Laub [L 7], [L 9], [L 10], Lee [L 16], Pappas et al [P 5], Van Dooren [V 3], [V 4]. It was observed by Laub [L 7] that the Hamiltonian and symplectic Schur forms can be obtained from the Schur form as follows:

14.1 THEOREM.

i) Let $\mathcal{H} \in \mathbf{C}^{m,m}$ be Hamiltonian having no purely imaginary eigenvalues and let $Q = \begin{bmatrix} Q_{11} & Q_{12} \\ Q_{21} & Q_{22} \end{bmatrix} \in U_{2n}(\mathbf{C})$ such that

$$(14.2) \qquad Q^*\mathcal{H}Q = \begin{bmatrix} T_{11} & T_{12} \\ 0 & T_{22} \end{bmatrix}$$

is in Schur form and $T_{11} \in \mathbf{C}^{n,n}$ has only stable eigenvalues. Let

$$(14.3) \qquad \tilde{Q} = \begin{bmatrix} Q_{11} & -Q_{21} \\ Q_{21} & Q_{11} \end{bmatrix}.$$

Then,

$$(14.4) \qquad \tilde{Q}^*\mathcal{H}\tilde{Q} = \begin{bmatrix} T_{11} & \tilde{T}_{12} \\ 0 & -T_{11}^* \end{bmatrix}$$

is in Hamiltonian Schur form and $\tilde{Q} \in US_{2n}(\mathbf{C})$.

ii) Let $S \in \mathbf{C}^{2n,2n}$ be symplectic having no eigenvalues on the unit circle and let $Q \in \begin{bmatrix} Q_{11} & Q_{12} \\ Q_{21} & Q_{22} \end{bmatrix} \in U_{2n}(\mathbf{C})$ such that

$$(14.5) \qquad Q^*SQ = \begin{bmatrix} T_{11} & T_{12} \\ 0 & T_{22} \end{bmatrix}$$

is in Schur form and $T_{11} \in C^{n,n}$ has only eigenvalues inside the unit circle. Let $\tilde{Q}$ be as in (14.3). Then,

$$(14.6) \qquad \tilde{Q}^* S \tilde{Q} = \begin{bmatrix} \tilde{T}_{11} & \tilde{T}_{12} \\ 0 & \tilde{T}_{11}^{-*} \end{bmatrix}$$

is in symplectic Schur form and $\tilde{Q} \in US_{2n}(C)$.

PROOF: From (14.3) we have

$$\begin{bmatrix} Q_{11} & -Q_{21} \\ Q_{21} & Q_{11} \end{bmatrix}^* \begin{bmatrix} Q_{11} & -Q_{21} \\ Q_{21} & Q_{11} \end{bmatrix} = \begin{bmatrix} Q_{11}^*Q_{11} + Q_{21}^*Q_{21} & -Q_{11}^*Q_{21} + Q_{21}^*Q_{11} \\ Q_{11}^*Q_{21} - Q_{21}^*Q_{11} & Q_{21}^*Q_{21} + Q_{11}^*Q_{11} \end{bmatrix}$$

and

$$\begin{bmatrix} Q_{11} & -Q_{21} \\ Q_{21} & Q_{11} \end{bmatrix}^* J \begin{bmatrix} Q_{11} & -Q_{21} \\ Q_{21} & Q_{11} \end{bmatrix} = \begin{bmatrix} Q_{11}^*Q_{21} - Q_{21}^*Q_{11} & Q_{11}^*Q_{11} + Q_{21}^*Q_{21} \\ -Q_{21}^*Q_{21} - Q_{11}^*Q_{11} & -Q_{21}^*Q_{11} + Q_{11}^*Q_{21} \end{bmatrix}.$$

Now $Q \in U_{2n}(C)$, thus $Q_{11}^*Q_{11} + Q_{21}^*Q_{21} = I$.

Now the columns of $\begin{bmatrix} Q_{11} \\ Q_{21} \end{bmatrix}$ span an invariant subspace, thus we have in i):

$$[-Q_{21}\ Q_{11}] M \begin{bmatrix} Q_{11} \\ Q_{21} \end{bmatrix} = [-Q_{21}\ Q_{11}] \begin{bmatrix} Q_{11} \\ Q_{21} \end{bmatrix} T_{11} = T_{11}^* [-Q_{21}\ Q_{11}] \begin{bmatrix} Q_{11} \\ Q_{21} \end{bmatrix}.$$

So the matrix $W = Q_{11}^*Q_{21} - Q_{21}^*Q_{11}$ solves the Lyapunov equation

$$(14.7) \qquad W T_{11} - T_{11}^* W = 0.$$

But T_{11} has only stable eigenvalues. Thus, it follows from the well–known theory for Lyapunov equations, e.g. Gantmacher [G 4], that $W = 0$.

In ii) we have

$$[-Q_{21}\ Q_{11}] M \begin{bmatrix} Q_{11} \\ Q_{21} \end{bmatrix} = [-Q_{21}\ Q_{11}] \begin{bmatrix} Q_{11} \\ Q_{21} \end{bmatrix} T_{11} = T_{11}^{-*} [-Q_{21}\ Q_{11}] \begin{bmatrix} Q_{11} \\ Q_{21} \end{bmatrix},$$

since M is symplectic. Thus, $W = Q_{21}^*Q_{21} - Q_{21}^*Q_{21}$ solves the Stein equation

$$(14.8) \qquad T_{11}^* W T_{11} - W = 0.$$

But T_{11} has only eigenvalues of modulus less than 1. Thus, by the well–known theory for Stein equation, e.g. Lancaster/Tismenetsky [L 5], we have $W = 0$. ∎

An analogous result can be obtained for symplectic pencils, e.g. Laub [L 12] and also in an obvious way for real Hamiltonian or symplectic matrices, using the real Schur form.

It follows that the QR- and QZ-algorithm, which compute the Schur form and generalized Schur form can be used to obtain the Hamiltonian and symplectic Schur forms. The QR- and QZ-algorithm are numerically backwards stable, e.g. Parlett [P 6], Wilkinson [W 8] or

Golub/Van Loan [G 12], so the resulting Schur form is the exact Schur form of a nearby matrix or pencil. The cost of applying these methods to Hamiltonian or symplectic matrices or symplectic pencils is approximately $200n^3$ for QR and $400n^3$ for QZ.

It has been pointed out, however, by several authors, Birdwell/Laub [B 15], Laub [L 7], [L 8], Kenney et al [K 15], Petkov et al [P 12], [P 13], [P 14], that the QR- and QZ-methods have to be combined with a proper scaling strategy and a balancing of the problem. A detailed dicussion for this is given in Kenney et al [K 15]. In this paper also it is shown under which condition Newton's method as defect correction method improves the results of the QR- and QZ-algorithm.

Unfortunately during the iteration process, both these algorithms ignore the Hamiltonian or symplectic structure, respectively. Thus, the given pairing of eigenvalues (see Section 4) is destroyed by roundoff errors and hence the methods can produce physically meaningless results. This can be fatal in particular, when eigenvalues close to the imaginary axis in the Hamiltonian case or close to the unit circle in the symplectic case exist, since then due to roundoff errors it is not possible to distinguish the required subspaces anymore, e.g. Van Loan [V 7], Petkov et al [P 12], [P 13], [P 14]. Also for Hamiltonian matrices with eigenvalues on the imaginary axis, for which still a Hamiltonian Schur form exists, e.g. Clements/Glover [C 8], they are not well suited. A modification of both methods was recently introduced for such problems in Clements/Glover [C 8], which describe a special eigenvector deflation technique that guarantees that the eigenvalues come out with the correct pairing. This is certainly an advantage over the general QR- or QZ-method but still this method is ignoring the structure in part during the process. We will briefly discuss this deflation technique here from a different viewpoint than that given in Clements/Glover [C 8].

Consider the usual QR-algorithm for $\mathcal{H} \in \mathbf{C}^{2n,2n}$ Hamiltonian. Let

$$(14.8) \qquad Q_0^* \mathcal{H} Q_0 \overset{\triangle}{=} \left[\begin{array}{c} \diagbox{} \end{array} \right]$$

and let at some stage of the iteration

$$(14.9) \qquad \mathcal{H}_i := Q_i^* \dots Q_0^* \mathcal{H} Q_0 \dots Q_i \overset{\triangle}{=} \left[\begin{array}{cc} \diagbox{} & | \\ & * \end{array} \right],$$

i.e. the last row of $\mathcal{H}_i$ is λe_n^*. Let $U = Q_0 \dots Q_i$ and λ be the last diagonal element of $\mathcal{H}_i$. Then $z^* = \begin{bmatrix} x \\ y \end{bmatrix}^* := e_{2n}^* U^*$ is a left eigenvector of $\mathcal{H}$ to the eigenvalue λ. Let $w = U^* J U e_{2n} = U^* J z$. Then w is right eigenvector of $\mathcal{H}_i$ to the eigenvalue $-\bar\lambda$, since

$$(14.10) \qquad \mathcal{H}_i w = \mathcal{H}_i U^* J z = U^* \mathcal{H} J z = -U^* J \mathcal{H}^* z = -\bar\lambda U^* J z = -\bar\lambda w.$$

Now $e_{2n}^* w = e_{2n}^* U^* \begin{bmatrix} -y \\ x \end{bmatrix} = [x^* y^*] \begin{bmatrix} -y \\ x \end{bmatrix} = -y^* x + x^* y$. In the real case this is trivially 0. in the complex case we have to use the fact that left and right eigenvectors of different eigenvalues are orthogonal. Thus, if $\lambda \neq -\bar\lambda$, which holds by assumption, then it follows that $e_{2n}^* w = 0$. Observe that also $\|w\|_2 = 1$. Let $V \in \mathcal{U}_{2n}(\mathbf{C})$ such that $V^* w = e_1$. Then the first column of V is w but also the last row and column of V is e_{2n}, so

$$(14.11) \qquad V^* \mathcal{H}_i V = \begin{bmatrix} -\bar\lambda & & * \\ 0 & \tilde{\mathcal{H}}_i & | \\ 0 & 0 & \lambda \end{bmatrix}$$

and we can deflate the first and last row and column of $V^*\mathcal{H}_iV$ and proceed with the smaller $\tilde{\mathcal{H}}_i$. The idea of Clements/Glover [C 8] is to compute V in a more efficient way by the following procedure:

14.12 ALGORITHM CCGDEF [CLEMENTS/GLOVER DEFLATION PROCEDURE].

Given a matrix $\mathcal{H}_i \in \mathbb{C}^{2n,2n}$ of the form (14.9) that is unitarily similar to a Hamiltonian matrix $\mathcal{H}$, i.e. $\mathcal{H}_i = U^*\mathcal{H}U$.

> Set $w = U^* JU \, e_{2n}$,
>
> FOR $j = 2n - 1, 2n - 2, \ldots, 2$
>
> > Let $G_j = $ CGIVGEN $(j, j + 1, w)$.
> >
> > Set $\mathcal{H}_i := G_j \, \mathcal{H}_i G_j^*$
> >
> > $U := UG_j^*$
> >
> > $W := G_j W$
>
> END FOR

END

At the end of this procedure we obtain that $\mathcal{H}_i$ is of the form (14.11) and it is possible to deflate rows and columns $1, 2n$ from the matrix $\mathcal{H}_i$ and proceed with a matrix of size $2n - 2$.

A problem that may occur here, is that the eigenvalue λ in the lower right corner has negative real part, which means that the deflation leaves them in the wrong order, so still an ordering of the eigenvalues has to follow the algorithm.

Essentially we can avoid that the wrong eigenvalues occur at the bottom by shifting always with eigenvalues with positive real part in QR or QZ. Then the probability that we will have to perform such an extra step is small. With these precautions the Clements/Glover deflation cuts the number of iterations per eigenvalue essentially into a half compared with the usual QR-algorithm and automatically produces the correct subspace. Similar constructions can be obtained in the symplectic case and also with 2×2 blocks and block elimination in the case of real matrices with complex eigenvalues. One can interpret this deflation procedure as one step of the RQ-algorithm applied to $H_i + \bar{\lambda}I$ after the eigenvalue λ has deflated. This shows the relationship to the shift strategy used in Byers [B 39], which in the explicit form is a QR-step applied to $H_i - \lambda I$ followed by an RQ-step applied to $H_i + \bar{\lambda}I$.

In order to retain the structure, for the continuous case, Paige/Van Loan [P 2] suggested to use a QR like algorithm based on unitary symplectic transformations to extract the relevant invariant subspace of a Hamiltonian matrix $\mathcal{H}$. In recent years, several variant QR or QZ like methods have been deleveloped, Bunse–Gerstner/Mehrmann [B 32], Bunse–Gerstner et al [B 36], Byers [B 39], [B 41], Mehrmann [M 6], but none of these methods is entirely satisfactory.

The following algorithm is an outline for an idealized QR-iteration for Hamiltonian matrices.

14.13 ALGORITHM [HAMILTONIAN QR FRAMWORK].

Given a Hamiltonian matrix $\mathcal{H} = \begin{bmatrix} F & G \\ H & -F^* \end{bmatrix} \in \mathbf{C}^{2n,2n}$, this algorithm computes a non-singular matrix $S \in \mathbf{C}^{2n,2n}$ such that $S^{-1}\mathcal{H}S = T$ is approximately Hamiltonian triangular (see Section 7).

Initialize $Q := I_{2n}$, $\mathcal{H}_0 = \mathcal{H}$.

FOR $k = 1, 2, 3, \ldots$ UNTIL T is "nearly" Hamiltonian triangular.

a) Select a holomorphic function $f_k(z)$ and factor $f(\mathcal{H}_{k-1}) = Q_k R_k$ for some unitary symplectic matrix $Q_k \in \mathbf{C}^{2n,2n}$ and some Hamiltonian triangular matrix $R_k \in \mathbf{C}^{2n,2n}$.

b) Set $T = \mathcal{H}_k = R_k Q_k$.

c) Set $Q := Q Q_k$.

END FOR

END

An easy induction shows that for all k, $\mathcal{H}Q = Q\mathcal{H}_k$, so Algorithm 14.13 is a sequence of unitary symplectic similarity transformations. In particular all the iterates are Hamiltonian.

Since a Hamiltonian triangular matrix is just a permutation of a usual triangular matrix, this method is a special case of the CQR iteration. It is known that under mild assumptions the iterates M_k tend to Hamiltonian triangular form as k goes to infinity, e.g. Parlett [P 6], Watkins [W 4], Wilkinson [W 8], or Golub/Van Loan [G 12].

The *shift function* $f_k(z)$ must be chosen in such a way that the structured QR-factorization in Step 1 a) exists. Moreover, to insure rapid convergence, it should be chosen so that $f_k(\mathcal{H}_{k-1})$ has a relatively large and/or a relatively small eigenvalue. A successful choice is the *Cayley shift*, Byers [B 39], [B 41].

$$(14.14) \qquad f_k(z) = (z - \lambda_k\, I)\, (z + \bar{\lambda}_k\, I)^{-1},$$

where λ_k is an approximate eigenvalue of $\mathcal{H}_{k-1}$. A typical choice is to take λ_k to be the $(2n, 2n)$ entry of $\mathcal{H}_{k-1}$. A consequence of the eigenvalue pairing property of Hamiltonian matrices (see Section 4) is that if λ is an eigenvalue of $\mathcal{H}_{k-1}$, then $\lambda_{k-1} - \lambda$ and $(\bar{\lambda}_{k-1} - \lambda)^{-1}$ are eigenvalues of $f_k(\mathcal{H}_{k-1})$. Thus, if λ_k approximates an eigenvalue of $\mathcal{H}_{k-1}$, then $f_k(\mathcal{H}_{k-1})$ has both a small eigenvalue and a large eigenvalue. (Strictly speaking if λ_k is an eigenvalue of $\mathcal{H}_{k-1}$, then $f_k(\mathcal{H}_{k-1})$ is undefined. However, the information that λ is an exact eigenvalue can be used to reduce the size of the problem. The algorithm never has to "invert" a singular matrix. See Bunch [B 22] or Byers [B 39] for details.) An easy calculation shows that for Hamiltonian $\mathcal{H}_k$, $f_k(\mathcal{H}_{k-1})$ is symplectic and the structured QR-factorization exists, Bunse–Gerstner [B 25], Byers [B 39], [B 41].

Another successful choice of shift function is

$$(14.15) \qquad f_k(z) = \frac{(z - \lambda_k)(z - \bar{\lambda}_k)}{(z + \bar{\lambda}_k)(z + \lambda_k)} = \frac{z^2 - 2\,\mathcal{R}e(\lambda_k) + |\lambda_k|^2}{z^2 + 2\,\mathcal{R}e(\lambda_k) + |\lambda_k|^2}$$

where again λ_k is an approximate eigenvalue of $\mathcal{H}_{k-1}$. With this shift function in the real case, the algorithm can avoid complex arithmetic, while still being able to take advantage of complex values of λ_k.

When carefully coded, (e.g. Golub/Van Loan [G 12]) this algorithm is numerically backwards stable, i.e. equivalent to perturbing the original Hamiltonian matrix $\mathcal{H}$. At completion, Q, T and $\mathcal{H}$ satisfy

$$(14.16) \qquad\qquad (\mathcal{H} + E)Q = QT$$

for some matrix $E \in \mathbf{R}^{2n, \times 2n}$ such that $\|E\| \leq \phi(n)\epsilon\|\mathcal{H}\|$, where $\phi(n)$ is a low degree polynomial in n and ϵ is the machine precision. If the transformations are unitary symplectic as described by Byers [B 39], [B 41], Mehrmann [M 6], Paige/Van Loan [P 2], see Section 13, then this method is even strongly backwards stable in the way defined by Bunch [B 22], e.g. in (14.16) the error matrix E is Hamiltonian. Nevertheless, it is still useful to follow the method by a few steps of defect correction.

The fatal disadvantage that Algorithm 14.12 has is that it uses an unordinate amount of work per iteration. It does not become practical until the work count is down to $\mathcal{O}(n^2)$ operations per iteration.

Following the classical work, introducing the QR-algorithm, Francis [F 4], [F 5], to reduce the work load, the iterates should be maintained in a Hessenberg–like condensed form. It was pointed out by Ammar [A 2], Ammar/Martin [A 3], Byers [B 39], that the natural form to use is the Hamiltonian Hessenberg form defined in Section 7. If $\mathcal{H}_0$ in Algorithm 14.13 is in this form and f_k is the Cayley shift (14.14) or (14.15) then all subsequent $\mathcal{H}'_k$s are also in Hamiltonian Hessenberg form. This form is uniquely determined (up to column scaling) by $\mathcal{H}_0$ and the first column of Q. Thus, $\mathcal{H}_k$ can be inferred from $\mathcal{H}_{k-1}$ and the first column of $f(\mathcal{H}_{k-1})$ without explicitly forming Q_k, R_k, or the remaining columns of $f(\mathcal{H}_k)$, see Byers [B 39]. So, the missing ingredient needed to make Algorithm 14.13 practical is a finite step algorithm to reduce the initial Hamiltonian matrix to Hamiltonian–Hessenberg form. Unfortunately, a suitable algorithm has been found only for the very special case of algebraic Riccati equations that come from single input control or single output control problems Byers [B 39], Mehrmann [M 6]. We will discuss these methods in detail in the next Section.

A reduction procedure, which is not quite sufficient but very helpful in the following was given by Paige/Van Loan [P 2]. They proposed a Hessenberg–like reduction method which

transforms a general matrix $M = \begin{bmatrix} M_{11} & M_{12} \\ M_{21} & M_{22} \end{bmatrix} \in \mathbf{C}^{2n,2n}$ to the form . This

algorithm is as follows:

14.17 ALGORITHM CHLRED [COMPLEX HESSENBERG LIKE REDUCTION ALGORITHM OF PAIGE/VAN LOAN].

Given a matrix $M = \begin{bmatrix} M_{11} & M_{12} \\ M_{21} & M_{22} \end{bmatrix} \in \mathbf{C}^{2n,2n}$ this algorithm determines $Q \in \mathcal{US}_{2n}(\mathbf{C})$ such that

$$Q^* M Q = \begin{bmatrix} \widetilde{M}_{11} & \widetilde{M}_{12} \\ \widetilde{M}_{21} & \widetilde{M}_{21} \end{bmatrix} \stackrel{\triangle}{=} \ \ \text{} \ .$$

FOR $k = 1, \ldots, n-1$

    IF $k \le n-2$ THEN

        Let $\begin{bmatrix} y \\ z \end{bmatrix} = M\ e_k$ and $Z_k = HHGEN(k+1, n, z)$.

        Set $M := \begin{bmatrix} Z_k^* & 0 \\ 0 & Z_k^* \end{bmatrix} M \begin{bmatrix} Z_k & 0 \\ 0 & Z_k \end{bmatrix}$.

    END IF

    Let $x = M\ e_k$ and $Q_k = GSGEN(k+1, x)$.

    Set $M := Q_k^* M Q_k$.

    IF $k \le n-2$ THEN

        Let $\begin{bmatrix} y \\ z \end{bmatrix} = M\ e_k$ and $\tilde{Z}_k = HHGEN(k+1, n, y)$.

        Set $M := \begin{bmatrix} \tilde{Z}_k^* & 0 \\ 0 & \tilde{Z}_k^* \end{bmatrix} M \begin{bmatrix} \tilde{Z}_k & 0 \\ 0 & \tilde{Z}_k \end{bmatrix}$.

    END IF

  END FOR

END

Applied to a Hamiltonian matrix, this algorithm produces again a Hamiltonian matrix, i.e.

the resulting matrix has the form $\begin{bmatrix} F & G \\ H & -F^* \end{bmatrix} \triangleq$.

Unfortunately this form is not reduced enough to serve as an initial reduction for Algorithm 14.13. Clearly there is also a real analogue of this method, which we call HLRED. The computational cost for this method is approximately $10n^3$ flops.

In a recent paper of Ammar/Mehrmann [A 4], it has been shown that a method that performs an initial reduction that stays invariant under the iteration of Algorithm 14.13, essentially solves the underdetermined system of quadratic equations

(14.18) $$z^T J \mathcal{H}^i z = 0 \quad i = 1, 3, 5, \ldots, 2n-1,$$

where z is the first column of the matrix Q that transforms $\mathcal{H}$ to Hamiltonian Hessenberg form. This property indicates why it is in general very hard to find this transformation and what the approximate amount of work will be.

A similar framework like that of Algorithm 14.13 can be given for symplectic martices and can be generalized for symplectic pencils. The analogous difficulties occur, plus an extra difficulty that does not occur in the Hamiltonian case. It is relatively simple to guarantee that all the iterates stay Hamiltonian even under the presence of roundoff errors, by using the symmetries in the Hamiltonian form. For symplectic matrices or pencils, this is not so easy, since the property is defined by an implicit equation. Thus, roundoff errors slowly destroy this property even if one works with unitary symplectic transformations, e.g. Flaschka et al [F 2]. Therefore, the perturbations due to roundoff do not preserve the structure, hence the method will not be strongly backwards stable. At this moment it is not known how to overcome this difficulty.

If the requirement that Algorithm 14.13 uses only unitary similarity transformations is relaxed, then other Hessenberg like condensed forms like the J–tridiagonal form or the symplectic Hessenberg pencil become available. For example the SR–algorithm of Bunse–Gerstner/Mehrmann [B 32] or Bunse–Gerstner et al [B 36] condense Hamiltonian matrices to J–tridiagonal form and the SZ-algorithm of Flaschka et al [F 2] condense real symplectic pencils to a symplectic Hessenberg pencil.

We shall discuss these methods in Section 16.

In this Section we have discussed a general framework for a QR like method for Hamiltonian matrices and shown what the numerical difficulties with this approach are. The most significant problem is the missing initial reduction to a Hessenberg like condensed form. This has only been achieved in special cases two of which we discuss in the following section. Other special cases with further algebraic strcuture are discussed in Bunse–Gerstner et al [B 28], [B 29].

§ 15 Unitary symplectic algorithms for special Hamiltonian or symplectic eigenvalue problems.

In Section 14 we have introduced a framework for a numerically strongly backwards stable method for the solution of the Hamiltonian eigenvalue problem. The missing part to make it a practical method is an initial reduction to Hamiltonian Hessenberg form. For the special case of Hamiltonian matrices arising from special single input or single output control problems such a reduction and a practical implementation of Algorithm 14.9 has been given by Byers [B 39], [B 41]. In this special case the Hamiltonian matrix

$$M = \begin{bmatrix} F & G \\ H & -F^* \end{bmatrix}$$

has a block H or G of rank 1, since the matrices C or B are of rank 1.

Analogously for the discrete case, i.e. for symplectic matrices or pencils arising from single input or single output control problems, a practical version of Algorithm 14.9 has been introduced by Mehrmann [M 6]. We discuss both methods in this section. Other special cases with further algebraic structure are discussed in a paper of Bunse–Gerstner et al [B 28], [B 29].

We first dicuss the Hamiltonian QR-algorithm of Byers [B 39], [B 41]. It consists of two steps, the initial reduction and the iteration.

15.1 ALGORITHM CHAMHES [COMPLEX UNITARY SYMPLECTIC REDUCTION OF A SPECIAL HAMILTONIAN MATRIX TO HAMILTONIAN HESSENBERG FORM].

Given a Hamiltonian matrix $\mathcal{H} \in \mathbf{C}^{2n,2n}$

$$\mathcal{H} = \begin{bmatrix} F & G \\ H & -F^* \end{bmatrix},$$

with $H = hh^*$, $h \in \mathbf{C}^n$, G Hermitian and $Q = I \in \mathbf{C}^{2n,2n}$, the following algorithm

overwrites $\mathcal{H}$ with the Hamiltonian Hessenberg matrix $Q^*\mathcal{H}Q \stackrel{\triangle}{=}$ $\begin{bmatrix} \diagdown & \square \\ & \diagdown \\ & * \end{bmatrix}$ where Q

is a product of Householder symplectic matrices.

Let

$$Z = \begin{bmatrix} \cdot^{,1} \\ {}_1\cdot^{,} \end{bmatrix} CHHGEN(1, n, h),$$

and set

$$\mathcal{H} := \begin{bmatrix} Z^* & 0 \\ 0 & Z^* \end{bmatrix} \mathcal{H} \begin{bmatrix} Z & 0 \\ 0 & Z \end{bmatrix}, \quad Q := Q \begin{bmatrix} Z & 0 \\ 0 & Z \end{bmatrix}.$$

FOR $\quad k = n, n-1, n-2, \ldots, 3$

Let $Z_k =$CHHGEN $(1, k-1, \overline{f}_{k,*})$, where $\overline{f}_{k,*}$ denotes the complex conjugate of the k-th row of F.

Set $\mathcal{H} := \begin{bmatrix} Z_k^* & 0 \\ 0 & Z_k^* \end{bmatrix} \mathcal{H} \begin{bmatrix} Z_k & 0 \\ 0 & Z_k \end{bmatrix}, \quad Q := Q \begin{bmatrix} Z_k & 0 \\ 0 & Z_k \end{bmatrix}.$

END FOR

END

Byers [B 41] proves that this transformation is unique up to a multiplication of Q with a unitary symplectic diagonal matrix and that this form stays invariant under the following implicit Hamiltonian QR step.

15.2 ALGORITHM CIMHQR [COMPLEX IMPLICIT HAMILTONIAN QR–STEP].

Given $\mathcal{H} \in C^{2n,2n}$ in unreduced Hamiltonian Hessenberg form and $Z \in US_{2n}(C)$, the following algorithm overwrites $\mathcal{H}$ with $Q^*\mathcal{H}Q$ and Z with ZQ, where Q is a product of Householder symplectic and Givens symplectic matrices and

$$Q^*q(\mathcal{H}) = T = \begin{bmatrix} T_{11} & T_{12} \\ 0 & T_{11}^{-*} \end{bmatrix} \triangleq \begin{bmatrix} \boxed{} & \boxed{} \\ & \boxed{} \end{bmatrix}.$$

Here $q(\mathcal{H}) = (\mathcal{H} + \mu I)(\mathcal{H} - \mu I)^{-1}$ is the Cayley shift (14.10) and μ is an approximation to an eigenvalue of $\mathcal{H}$.

Let $\mathcal{H} = \begin{bmatrix} F & G \\ H & -F^* \end{bmatrix} \triangleq \begin{bmatrix} \boxed{} & \boxed{} \\ * & \boxed{} \end{bmatrix}.$

Let $x = (F + \mu I)e_1$.

FOR $\quad k = 1, \dots, n-1$

Let $Q_k = \text{CHHGEN } (k, k+1, x)$.

Set $F := Q_k^* F Q_k$, $G := Q_k^* G Q_k$, $H = Q_k^* H Q_k$, $Z := Z \begin{bmatrix} Q_k & 0 \\ 0 & Q_k \end{bmatrix}$, $x := F e_k$.

END FOR

(The matrices then are of the forms:

$$F \triangleq \begin{bmatrix} \boxed{} \end{bmatrix}, \ G \triangleq \begin{bmatrix} \boxed{} \end{bmatrix}, \ H \triangleq \begin{bmatrix} & \\ \cdot\cdot & \end{bmatrix}.)$$

Let $x \in C^{2n}$ with $x_n = 2[Re(f_{n,n}h_{n-1,n-1}) + Re(h_{n,n-1}\overline{f}_{n-1,n})]$, $x_{2n} = h_{n-1,n-1}g_{n,n} + |f_{n-1,n}|^2$ and let $P = \text{CGSGEN } (n, x)$. (Observe that P is real since x_n, x_{2n} are real.)

Let $\mathcal{H} := P^* \mathcal{H}P =: \begin{bmatrix} F & G \\ H & -F^* \end{bmatrix} \triangleq \begin{bmatrix} \boxed{} & \boxed{} \\ & \boxed{} \end{bmatrix}, \ Z := ZP$.

Let k be the index of the row of $\mathcal{H}$ with maximal norm, $(k = n$ or $k = n-1)$.

Let $x = \mathcal{H} e_k$ and let $Q_n = \text{CHHGEN } (n-1, n, x)$.

Set $F := Q_n^* F Q_n$, $G := Q_n^* G Q_n$, $H := Q_n^* H Q_n$, $Z := Z \begin{bmatrix} Q_n & 0 \\ 0 & Q_n \end{bmatrix}$

FOR $k = n, n-1, \dots, 3$

Let $x = \begin{bmatrix} 1 \\ & \cdot^{\cdot^{\cdot}} \end{bmatrix} F^* e_k$, $Z_k = \begin{bmatrix} 1 \\ & \cdot^{\cdot^{\cdot}} \end{bmatrix}$ CHHGEN $(n-k-2, n-k-1, x)$, such that

$$
Z_k^* \begin{bmatrix} 0 \\ \vdots \\ 0 \\ \overline{f}_{k,k-2} \\ \overline{f}_{k,k-1} \\ \vdots \\ \overline{f}_{k,n} \end{bmatrix} \overset{\wedge}{=} \begin{bmatrix} 0 \\ \vdots \\ 0 \\ 0 \\ * \\ \vdots \\ * \end{bmatrix} \Bigg\} k-2 \quad .
$$

Set $\mathcal{H} := \begin{bmatrix} Z_k^* & 0 \\ 0 & Z_k^* \end{bmatrix} \begin{bmatrix} F & G \\ H & -F^* \end{bmatrix} \begin{bmatrix} Z_k & 0 \\ 0 & Z_k \end{bmatrix}$, $Z := Z \begin{bmatrix} Z_k & 0 \\ 0 & Z_k \end{bmatrix}$.

END FOR

END

In the real case we have the analogous procedures HAMHES, IMHQR, where in order to avoid complex arithmetic, for nonreal shifts μ, we have to perform an even more complicated step by chosing the shift (14.11).

(15.3) $$ q(\mathcal{H}) = (\mathcal{H} + \mu I)(\mathcal{H} + \overline{\mu} I)[(\mathcal{H} - \mu I)(\mathcal{H} - \overline{\mu} I)]^{-1}. $$

We will not describe this step here, it is described in detail in Byers [B 39] or [B 41].

Based on these two steps we have the following algorithm to compute the Hamiltonian Schur-form.

15.4 ALGORITHM CHAMQR [COMPLEX HAMILTONIAN QR ALGORITHM].

Given a Hamiltonian matrix
$$ \mathcal{H} = \begin{bmatrix} F & G \\ H & -F^* \end{bmatrix} $$

having no eigenvalues with real part zero, and a tolerance ϵ, the following algorithm computes $Q \in \mathcal{US}_{2n}(\mathbb{C})$ such that

$$ Q^* \mathcal{H} Q \overset{\wedge}{=} \begin{bmatrix} \searrow & \square \\ & \searrow \end{bmatrix} $$

is in Hamiltonian Schur-form.

Step 1: Apply Algorithm 15.1 CHAMHES to compute $Q \in \mathcal{US}_{2n}(\mathbb{C})$ such that

$$ \begin{bmatrix} F & G \\ H & -F^* \end{bmatrix} := Q^* \mathcal{H} Q \overset{\wedge}{=} \begin{bmatrix} \searrow & \square \\ * & \searrow \end{bmatrix}. $$

Step 2:

    FOR $j = 1, 2 \ldots$

        Set all subdiagonal elements of F satisfying

(15.5)
$$|f_{i,i-1}| \le \epsilon(|f_{i,i}| + |f_{i-1,i-1}|)$$

        to zero. Find the largest $q \ge 0$ such that

$$\mathcal{H} = \begin{bmatrix} F_{11} & F_{12} & G_{11} & G_{12} \\ 0 & F_{22} & G_{21} & G_{22} \\ 0 & 0 & -F_{11}^* & 0 \\ 0 & H_{22} & -F_{12}^* & -F_{22}^* \end{bmatrix} \begin{matrix} \}q \\ \}n-q \\ \}q \\ \}n-q \end{matrix} \quad ,$$

        where F_{22} is unreduced upper Hessenberg.

        IF $q = n - 1$ THEN triangularize

(15.6)
$$\begin{bmatrix} I & 0 & 0 & 0 \\ 0 & F_{22} & 0 & G_{22} \\ 0 & 0 & I & 0 \\ 0 & H_{22} & 0 & -F_{22}^* \end{bmatrix}$$

        by applying a symplectic Givens transformation $Q_1 \in \mathcal{US}_{2n}(\mathbf{C})$ that elimi-
nates H_{22}.

        Set $\mathcal{H} := Q_1^* \mathcal{H} Q_1$, $Q := QQ_1$.

        Apply QR to F_{11} to compute $Q_2 \in \mathcal{U}_{n-1}(\mathbf{C})$ such that $Q_2^* F Q_2 \triangleq \begin{bmatrix} \diagdown \end{bmatrix}$.
Set $\tilde{Q}_2 = \text{diag}(Q_2, 1, Q_2, 1)$ and

$$\mathcal{H} := \tilde{Q}_2^* \mathcal{H} \tilde{Q}_2, \ Q := Q \tilde{Q}_2.$$

        QUIT

        END IF

        Set $\tilde{\mathcal{H}} = \begin{bmatrix} F_{22} & G_{22} \\ H_{22} & -F_{22}^* \end{bmatrix}$, $\tilde{Z} = I \in \mathbf{C}^{n-q,n-q}$ and apply Algorithm 15.3 CIMHQR
to $\tilde{\mathcal{H}}, \tilde{Z}$.

        Let $Z' := \begin{bmatrix} I & 0 & 0 & 0 \\ 0 & \tilde{Z}_{11} & 0 & \tilde{Z}_{12} \\ 0 & 0 & I & 0 \\ 0 & \tilde{Z}_{21} & 0 & \tilde{Z}_{22} \end{bmatrix}$ where $\tilde{Z}_{11}, \tilde{Z}_{12}, \tilde{Z}_{21}, \tilde{Z}_{22}$ are the blocks of $\tilde{Z}$.

        Set $\mathcal{H} = Z'^* \mathcal{H} Z'$, $Q := QZ'$.

    END FOR

Step 3:

Interchange the eigenvalues by the procedure described in Byers [B 39], which is essentially the same procedure as in the general QR-algorithm, i.e. compute $Q_0 \in \mathcal{U}S_{2n}(\complement)$ such that

$$\mathcal{H} := \begin{bmatrix} F & G \\ 0 & -F^* \end{bmatrix} := Q_3^* \mathcal{H} Q_3 \triangleq \begin{bmatrix} \searrow & \square \\ & \searrow \end{bmatrix}$$

and F has only eigenvalues with negative real part.

Set $Q := QQ_3$.

END

In the real case, where we have the slightly more general assumption that no nonzero eigenvalues with real part zero exist, the algorithm is called HAMQR, it uses HAMHES, IMHQR and the corresponding real generators of Householder symplectic and Givens symplectic matrices.

The final form then is $\mathcal{H} = \begin{bmatrix} T_{11} & T_{12} \\ 0 & -T_{11}^T \end{bmatrix}$, where T_{11} is quasi upper triangular and has no eigenvalues with negative real part. Observe that the matrix in (15.6), by the assumption that no nonzero eigenvalues with zero real part exists, can have only real eigenvalues in the real case. For the shift μ, one usually takes the first diagonal element or the eigenvalues $\mu, \bar{\mu}$ of the leading 2×2 principal submatrix of $\mathcal{H}$. For further details on this algorithm see Byers [B 39], [B 41].

Since this algorithm is exactly of the form discussed in Algorithm 14.9, it is numerically strongly backward stable. Observe that if G has rank 1 then we can exchange the roles of G and H by a block permutation. The computational cost for this algorithm is approximately $40n^3$ for the complete algorithm using the estimated number of iterations needed to be one iteration per eigenvalue.

The analogous algorithm for symplectic matrices and symplectic pencils, coming from single input or single output systems was introduced by Mehrmann [M 6]. Although there are essentially two algorithms, one for pencils and one for matrices, we will restrict ourselves mainly to the matrix case and reduce the pencil case to it. The first part is essentially the same for both algorithms. We begin with the description of this procedure for the pencil case and we discuss only the case that H has rank 1. Again we may exchange the roles of G, H if G has rank 1.

15.7 ALGORITHM CSPHESP [COMPLEX UNITARY SYMPLECTIC REDUCTION OF A SPECIAL SYMPLECTIC PENCIL TO A SYMPLECTIC HESSENBERG PENCIL].

Given a symplectic pencil

$$\alpha \mathcal{L}' - \beta \mathcal{M}' = \alpha \begin{bmatrix} F & 0 \\ H & I \end{bmatrix} - \beta \begin{bmatrix} I & -G \\ 0 & F^* \end{bmatrix},$$

where $H = thh^*$, $h \in \mathbb{C}^n$, $t \in \mathbb{R}$, the following algorithm overwrites $\alpha \mathcal{L}' - \beta \mathcal{M}'$ with

$$\alpha Q^* \mathcal{L}' Q - \beta Q^* \mathcal{M}' Q \triangleq \alpha \begin{bmatrix} \boxed{} \\ * \end{bmatrix} - \beta \begin{bmatrix} \\ \boxed{} \end{bmatrix},$$

where Q is a product of Householder symplectic matrices.

Let $Z = [\begin{smallmatrix} 1 \\ & \ddots \end{smallmatrix}]$ CHHGEN $(1, n, h)$, i.e. $Z^* h = \alpha e_n$ and set

$$F := Z^* F Z, \ H := Z^* H Z, \ G := Z^* G Z, \ Q := \begin{bmatrix} Z & 0 \\ 0 & Z \end{bmatrix}.$$

FOR $k = n, n-1, \ldots, 3$

Let $Z_k = CHHGEN(1, k-1, \overline{f}_{k,*})$, where $\overline{f}_{k,*}$ denotes the complex conjugate of the k–th row of F.

Set $F := Z_k^* F Z_k$, $G := Z_k^* G Z_k$, $Q := Q \begin{bmatrix} Z_k & 0 \\ 0 & Z_k \end{bmatrix}$.

END FOR

END

Then $\alpha \begin{bmatrix} F & 0 \\ H & I \end{bmatrix} - \beta \begin{bmatrix} I & -G \\ 0 & F^* \end{bmatrix} \triangleq \alpha \begin{bmatrix} \boxed{} \\ * \end{bmatrix} - \beta \begin{bmatrix} \\ \boxed{} \end{bmatrix}.$

If we have given a symplectic matrix

$$S = \begin{bmatrix} S_{11} & S_{12} \\ S_{21} & S_{22} \end{bmatrix},$$

where rank $S_{21} = 1$, i.e. $S_{21} = ph^*$, then we first reduce h as above and then reduce the matrix S_{11} the same way we reduced F. Observe that after reducing h, the blocks S_{11} and F differ only in the last column, which is not touched by the reduction procedure. In this case the corresponding procedure is called CSPHES and the final form is

(15.8) $\qquad S = [S_{ij}] = \begin{bmatrix} S_{11} & S_{12} \\ S_{21} & S_{22} \end{bmatrix} \triangleq \begin{bmatrix} \boxed{} & \boxed{} \\ | & \boxed{} \end{bmatrix}.$

Observe that it is very easy to obtain the blocks of the corresponding pencil $\alpha \mathcal{L}' - \beta \mathcal{M}'$ from S, using the following simple result:

15.9 PROPOSITION [THREE FACTOR THEOREM].

Let $S \in S_{2n}(\mathbb{C})$, $S = \begin{bmatrix} S_{11} & S_{12} \\ S_{21} & S_{22} \end{bmatrix}$. If S_{11}^{-1} exists, then S has a factorization

(15.10) $\qquad S = \begin{bmatrix} I & 0 \\ -H & I \end{bmatrix} \begin{bmatrix} F^{-1} & 0 \\ 0 & F^* \end{bmatrix} \begin{bmatrix} I & -G \\ 0 & I \end{bmatrix}$

and if S_{22}^{-1} exists, S has a factorization

(15.11)
$$S = \begin{bmatrix} I & G \\ 0 & I \end{bmatrix} \begin{bmatrix} F & 0 \\ 0 & F^{-*} \end{bmatrix} \begin{bmatrix} I & 0 \\ H & I \end{bmatrix},$$

where $G = G^*$, $H = H^*$.

PROOF: If S_{11} is invertible then

$$S = \begin{bmatrix} I & 0 \\ S_{21}S_{11}^{-1} & I \end{bmatrix} \begin{bmatrix} S_{11} & 0 \\ 0 & S_{11}^{-*} \end{bmatrix} \begin{bmatrix} I & S_{11}^{-1}S_{12} \\ 0 & I \end{bmatrix}$$

where the equation

$$S_{22} = S_{21}S_{11}^{-1}S_{12} + S_{11}^{-*}$$

follows from $S^*JS = J$.

Analogously we obtain the other factorization if S_{22} is invertible. ∎

Thus, using (15.11) we obtain in the reverse direction

(15.12)
$$S_{11} = F + GF^{-*}H, \quad S_{21} = F^{-*}H, \quad S_{22} = F^{-*}, \quad S_{12} = GF^{-*}.$$

If S is in symplectic Hessenberg form, S_{11} is unreduced, and $H = thh^*$ then F is unreduced and it follows that S_{22} has nonzero elements only in the last column. For $H = \gamma\, e_n e_n^*$ we then get the formulas:

(15.13)
$$\gamma = s_{j,n}/s_{j,2n} \text{ for some } j \in \{1,\dots,n\} \text{ for which } s_{j,2n} \neq 0.$$

(15.14)
$$F = S_{11} - \gamma S_{12} e_n e_n^*.$$

(15.15)
$$G = S_{12} F^*.$$

In the following we will discuss only the case that F is nonsingular. This can be guaranteed by a deflation procedure, which deflates the parts of S_{11}, F corresponding to zero eigenvalues of S_{22} or in the pencil case, the zero and infinite eigenvalues of

$$\alpha \mathcal{L}' - \beta \mathcal{M}' = \alpha \begin{bmatrix} F & 0 \\ H & I \end{bmatrix} - \beta \begin{bmatrix} I & -G \\ 0 & F^* \end{bmatrix}.$$

This procedure was introduced for the special symplectic pencils discussed here in Mehrmann [M 6] and for general symplectic pencils in Flaschka et al [F 2].

15.16 ALGORITHM CSP0DEF [COMPLEX SYMPLECTIC DEFLATION OF 0 AND INFINITE EIGENVALUES].

Given a symplectic pencil

$$\alpha \mathcal{L}' - \beta \mathcal{M}' = \alpha \begin{bmatrix} F & 0 \\ H & I \end{bmatrix} - \beta \begin{bmatrix} I & -G \\ 0 & F^* \end{bmatrix},$$

this algorithm determines a symplectic matrix T and an invertible matrix $\tilde{T}$ such that

$$(15.18) \quad \tilde{T}(\mathcal{L}' - \lambda\mathcal{M}')T = \alpha \overbrace{\begin{bmatrix} 0 & \tilde{F}_1 & 0 & 0 \\ 0 & \tilde{F} & 0 & 0 \\ 0 & 0 & I & 0 \\ 0 & \tilde{H} & 0 & I \end{bmatrix}}^{\substack{n-k \quad k}} - \substack{n-k\{ \\ k\{} \begin{bmatrix} I & 0 & -\tilde{G}_{11} & -\tilde{G}_{12} \\ 0 & I & -\tilde{G}_{21} & -\tilde{G}_{22} \\ 0 & 0 & 0 & 0 \\ 0 & 0 & \tilde{F}_1^* & \tilde{F}^* \end{bmatrix},$$

with $\tilde{H} = \tilde{H}^*$, $\tilde{G} = \tilde{G}^*$, $\tilde{F}$ nonsingular.

Step 1:

Use the QR-decomposition with column pivoting, e.g. Golub/Van Loan [G 12] to determine $Q \in C^{n,n}$ unitary, $R \in C^{n,n}$ upper triangular and a permutation matrix P such that

$$(15.19) \qquad\qquad PF = RQ =: [0, R_2] \begin{bmatrix} Q_1 \\ Q_2 \end{bmatrix}$$

and R_2, Q_2 are of full rank.

Step 2: Set

$$T_1 := \begin{bmatrix} I & 0 \\ -H & I \end{bmatrix}, \ T_2 := \begin{bmatrix} Q^* & 0 \\ 0 & Q^* \end{bmatrix}, \ T_3 := \begin{bmatrix} P & 0 \\ 0 & Q \end{bmatrix},$$

$$T_4 := \begin{bmatrix} Q(I+GH)^{-1}P^* & 0 \\ 0 & I \end{bmatrix}, \ T_5 := \begin{bmatrix} I & 0 \\ R^*PHQ^* & I \end{bmatrix}$$

$$T := T_1 T_2, \ \tilde{T} = T_5 T_4 T_3$$

and form

$$
\begin{aligned}
&\alpha\tilde{\mathcal{L}}' - \beta\widetilde{\mathcal{M}}' := \tilde{T}(\alpha\mathcal{L}' - \beta\mathcal{M}')T = \\
(15.20) \quad &\alpha \begin{bmatrix} Q(I+GH)^{-1}P^*R & 0 \\ R^*P(H(I+GH)^{-1}P^*R & I \end{bmatrix} - \beta \begin{bmatrix} I & -Q(I+GH)^{-1}GQ^* \\ 0 & R^*P(-H(I+GH)^*Q^* \end{bmatrix},
\end{aligned}
$$

END

Then $\alpha\tilde{\mathcal{L}}' - \beta\widetilde{\mathcal{M}}'$ has the required form and the first $n-k$ columns of T span the right deflating subspace of $\alpha\mathcal{L}' - \beta\mathcal{M}'$ to the eigenvalues 0 and the rows $n+1,\ldots,2n-k$ of $\tilde{T}$ span the left deflating subspace to the eigenvalues ∞. We may therefore delete rows and columns $1,\ldots,n-k$, $n+1,\ldots,2n-k$ and proceed with the reduced pencil

$$(15.21) \qquad\qquad \hat{\mathcal{L}}' - \lambda\widehat{\mathcal{M}}' = \begin{bmatrix} \tilde{F} & 0 \\ \tilde{H} & I \end{bmatrix} - \lambda \begin{bmatrix} I & -\tilde{G} \\ 0 & \tilde{F}^* \end{bmatrix}.$$

All the computations in this algorithm can be carried out in numerically reliable way. The solution of linear systems with $I + GH$ is well–conditioned, in the case we consider here, coming from control problems, since then G, H are positive semidefinite.

In the special case discussed in this Section where $H = thh^*$ is rank 1 we even have by the Sherman/Morrison formula, e.g. Golub/Van Loan [G 12]

$$(15.22) \qquad (I + GH)^{-1} = (I + Gthh^*)^{-1} = I - \frac{1}{1 + th^*Gh} Gthh^*.$$

In the real case we have an analogous procedure which we call SPODEF.

It is shown in Mehrmann [M 6] that the symplectic Hessenberg form stays invariant under the following iteration step for symplectic matrices. Assume that S_{11} is unreduced.

15.23 ALGORITHM CIMSPQR [COMPLEX IMPLICIT SYMPLECTIC QR–STEP].

Given $S = \begin{bmatrix} S_{11} & S_{12} \\ S_{21} & S_{22} \end{bmatrix} \in \mathcal{S}_{2n}(\mathbf{C})$ in symplectic Hessenberg form, i.e. $S \triangleq \begin{bmatrix} \boxed{} & \boxed{} \\ | & \boxed{} \end{bmatrix}$,

with S_{11} unreduced, and $Z \in \mathcal{US}_{2n}(\mathbf{C})$. Assume that S_{11} is nonsingular. Let

$$s_{j,2n} = \max_{i \in \{n+1, \dots, 2n\}} \{|s_{i,2n}|\}, \; \gamma = s_{j,n}/s_{j,2n}, \; F = S_{11} - \gamma S_{12}e_n e_n^*, \; G = S_{12}F, \; H = \lambda_n e_n e_n^*.$$

The following algorithm overwrites S with $Q^*SQ \triangleq \begin{bmatrix} \boxed{} & \boxed{} \\ | & \boxed{} \end{bmatrix}$ and Z with ZQ, where

$Q \in \mathcal{US}_{2n}(\mathbf{C})$ is a product of Householder symplectic and Givens symplectic matrices and

$Q^*q(S) \triangleq \begin{bmatrix} \boxed{} & \boxed{} \\ & \boxed{} \end{bmatrix}$, where

$$(15.24) \qquad q(S) = (S - \lambda I)^{-1}(\bar{\lambda}S - I).$$

Let $\tilde{F} := \bar{\lambda}F - I$, $y_{2n} := \bar{\lambda}\gamma$, $x = (\bar{\lambda}S_{11} - I)e_n$, $v = e_n$.

FOR $k = 1, \dots, n-1$

Let $Z_k = $ CHHGEN $(k, k+1, \tilde{f}_{*,k})$ and set

$$S := \begin{bmatrix} Z_k^* & 0 \\ 0 & Z_k^* \end{bmatrix} S \begin{bmatrix} Z_k & 0 \\ 0 & Z_k \end{bmatrix}, \; Z := Z \begin{bmatrix} Z_k & 0 \\ 0 & Z_k \end{bmatrix}, \; v := Z_k v.$$

END FOR

Let $y_n = v^*x$ and let y be arbitrary, with the given components y_{2n}, y_n.

Let $U_1 = $ CGSGEN (n, y) and let c_1, s_1 be the parameters determining U_1.

Set

$$w_{2n} := -s_1 e_n^* S_{12} e_n + c_1 e_n^* (I - \lambda F) v,$$
$$w_n := -s_1 e_n^* (S_{11} - \lambda I) e_n + c_1 \gamma e_n^* v$$

and determine c_2, s_2 by applying Algorithm 13.4 to determine $U_2 \in \mathcal{U}S_{2n}(\mathbb{C})$ such that $w^* U_2$ has a zero n–th component. Set

$$Z := Z U_2, \quad S = U_2^* S U_2 = \begin{bmatrix} S_{11} & S_{12} \\ S_{21} & S_{22} \end{bmatrix} \overset{\wedge}{=} \begin{bmatrix} \boxed{} & \boxed{} \\ | & \boxed{} \end{bmatrix}$$

S_{21} is again rank 1, i.e. $S_{21} = p h^*$.

Let $U_3 = \begin{bmatrix} 1 & & & & \\ & \ddots & & & \\ & & 1 & & \\ & & & 0 & 1 \\ & & & 1 & 0 \end{bmatrix}$ CHHGEN $(n-1, n, h)$ and set

$$S := \begin{bmatrix} U_3^* & 0 \\ 0 & U_3^* \end{bmatrix} S \begin{bmatrix} U_3 & 0 \\ 0 & U_3 \end{bmatrix}, \quad Z := Z \begin{bmatrix} U_3 & 0 \\ 0 & U_3 \end{bmatrix}.$$

Let $x = e_n^* S_{11}$.

FOR $k = n, n-1, \ldots, 3$

    Let $Z_k = $ CHHGEN $(k-2, k-1, x)$.

    Set $S = \begin{bmatrix} Z_k^* & 0 \\ 0 & Z_k^* \end{bmatrix} S \begin{bmatrix} Z_k & 0 \\ 0 & Z_k \end{bmatrix}, \quad Z := Z \begin{bmatrix} Z_k & 0 \\ 0 & Z_k \end{bmatrix}.$

END FOR

END

In the real case we have the analogous procedures SPHESP, SPHES, SP0DEF, IMSPQR, where in order to avoid complex arithmetic, for nonreal shifts λ we take

(15.25) $$q(S) = (S - \lambda I)^{-1} (S - \bar{\lambda} I)^{-1} (\bar{\lambda} S - I)(\lambda S - I).$$

We do not describe this algorithm in detail here but refer to Mehrmann [M 6] for a description of this part.

Based on these steps we then have the following algorithm to compute the symplectic Schur–form.

15.26 ALGORITHM CSPQR [COMPLEX SYMPLECTIC QR ALGORITHM]..

Given $S \in \mathcal{S}_{2n}(\mathbb{C})$, $S = \begin{bmatrix} S_{11} & S_{12} \\ S_{21} & S_{22} \end{bmatrix}$ with rank $S_{21} = 1$ having no eigenvalues of modulus 1 and given a tolerance ϵ, the following algorithm computes $Q \in \mathcal{U}S_{2n}(\mathbb{C})$ such that

$$Q^* S Q \overset{\wedge}{=} \begin{bmatrix} \boxed{} & \boxed{} \\ & \boxed{} \end{bmatrix}$$

is in symplectic Schur–form.

Step 1:

Apply Algorithm 15.8 CSPHES to compute $Q \in \mathcal{US}_{2n}(\mathbf{C})$ such that

$$Q^*SQ = \begin{bmatrix} S_{11} & S_{12} \\ S_{21} & S_{22} \end{bmatrix} \overset{\triangle}{=} \begin{bmatrix} \boxed{\diagdown} & \square \\ \vert & \square \end{bmatrix}$$

is in symplectic Hessenberg form.

Step 2:

FOR $i = 1, 2, \ldots$

Set all subdiagonal elements of S_{11} satisfying

(15.27)
$$|s_{i,i-1}| \leq \epsilon(|s_{i,i}| + |s_{i-1,i-1}|)$$

to zero. Find the largest $q \geq 0$ such that

(15.28)
$$S = \begin{bmatrix} \Sigma_{11} & \Sigma_{12} & \Sigma_{13} & \Sigma_{14} \\ 0 & \Sigma_{22} & \Sigma_{23} & \Sigma_{24} \\ 0 & \Sigma_{32} & \Sigma_{33} & \Sigma_{34} \\ 0 & \Sigma_{42} & \Sigma_{43} & \Sigma_{44} \end{bmatrix} \begin{matrix} \}q \\ \}n-q \\ \}q \\ \}n-q \end{matrix}$$

and such that Σ_{22} is unreduced upper Hessenberg. (Σ_{32} is 0 then, since $S \in S_{2n}(\mathbf{C})$).

IF $q = n - 1$ THEN triangularize

(15.29)
$$\begin{bmatrix} I & 0 & 0 & 0 \\ 0 & \Sigma_{22} & 0 & \Sigma_{24} \\ 0 & 0 & I & 0 \\ 0 & \Sigma_{42} & 0 & \Sigma_{44} \end{bmatrix}$$

by applying a symplectic Givens transformation $Q_1 \in \mathcal{US}_{2n}(\mathbf{C})$, that eliminates Σ_{42}.

Set $S := Q_1 S Q_1^*$, $Q := Q Q_1$.

Apply the standard QR-algorithm to Σ_{11} and compute $Q_2 \in \mathcal{U}_{n-1}(\mathbf{C})$ such that $Q_2^* \Sigma_{11} Q_2 \overset{\triangle}{=} \begin{bmatrix} \diagdown \end{bmatrix}$.

Set $\tilde{Q}_2 = \text{diag}(Q_2, 1, Q_2, 1)$ and $S := \tilde{Q}_2^* S \tilde{Q}_2$, $Q := Q \tilde{Q}_2$.

QUIT

END IF

Set $\tilde{S} = \begin{bmatrix} \Sigma_{22} & \Sigma_{24} \\ \Sigma_{42} & \Sigma_{44} \end{bmatrix}$, $\tilde{Z} = I \in \mathbb{C}^{n-q,n-q}$.

Apply Algorithm 5.27 CSP0DEF to deflate zero eigenvalues and then apply Algorithm 15.23 CIMSPQR to $\tilde{S}, \tilde{Z}$.

Let $Z' = \begin{bmatrix} I & 0 & 0 & 0 \\ 0 & \tilde{Z}_{11} & 0 & \tilde{Z}_{12} \\ 0 & 0 & I & 0 \\ 0 & \tilde{Z}_{21} & 0 & \tilde{Z}_{22} \end{bmatrix}$, where $\tilde{Z}_{11}, \tilde{Z}_{12}, \tilde{Z}_{21}, \tilde{Z}_{22}$ are the blocks of $\tilde{Z}$.

Set $S := Z'^* S Z'$, $Q := QZ'$.

END FOR

Step 3:

The ordering of the eigenvalues, such that all eigenvalues with modulus less than 1 are in the top left part is obtained by a similar procedure as in the Hamiltonian case. (E.g. Mehrmann [M 6]), i.e. we compute $Q_3 \in \mathcal{U}S_{2n}(\mathbb{C})$ such that

$$S = \begin{bmatrix} T_{11} & T_{12} \\ 0 & T_{11}^{-*} \end{bmatrix} := Q_3^* S Q_3 \overset{\Delta}{=} \begin{bmatrix} \text{figure} \end{bmatrix}$$

and T_{11} has only eigenvalues of modulus less than 1. Set $Q := QQ_3$.

END

In the real case, the algorithm is called SPQR, it uses SPHES, IMSPQR and the corresponding generators for Householder symplectic and Givens symplectic matrices.

The final form is $S = \begin{bmatrix} T_{11} & T_{12} \\ 0 & T_{11}^{-T} \end{bmatrix}$, where T_{11} is quasi upper triangular and has no eigenvalues of modulus greater or equal 1. As shifts we usually take the top diagonal element of the eigenvalues of the leading 2×2 principal submatrix of S.

Note that if we begin with a symplectic pencil in symplectic Hessenberg form then it is possible to replace CIMSPQR, CSPQR by QZ variants. But in Algorithm 15.23 we have to compute y_n, w_{2n}, w_n to determine the matrices U_1, U_2. In order to do this, we need parts of

$$S = \begin{bmatrix} I & -G \\ 0 & F^* \end{bmatrix}^{-1} \begin{bmatrix} F & 0 \\ H & I \end{bmatrix}.$$

By applying first Algorithm 15.27 CSP0DEF to deflate $0, \infty$ eigenvalues we can guarantee that the part of F that we work with is nonsingular.

We can then determine the matrix that transforms the pencil from the right similar as in Algorithm 15.23 and then determine a nonunitary transformation from the left, that transforms then pencil back in symplectic Hessenberg form. But it is not always guaranteed that this left transformation can be computed in a stable way at intermediate iteration steps and also we need solutions of

$$F^* x = v$$

to determine the right transformation matrices. From this point of view it seems more reasonable to actually compute S as in (15.29), using the QR-decomposition of the non-singular part of F and then applying the previous algorithm. So we have the following procedure in the pencil case.

15.30 ALGORITHM CSPQRP [COMPLEX SYMPLECTIC QR-ALGORITHM FOR SYMPLECTIC PENCILS].

Given a symplectic pencil having no eigenvalues of modulus 1,

$$\alpha \mathcal{L}' - \beta \mathcal{M}' = \alpha \begin{bmatrix} F & 0 \\ H & I \end{bmatrix} - \beta \begin{bmatrix} I & -G \\ 0 & F^* \end{bmatrix}$$

with rank $H = 1$, G, H positive semidefinite F nonsingular and given a tolerance ϵ, the following algorithm computes $Z \in \mathcal{S}_{2n}(\mathbb{C})$ and S invertible such that

$$S(\alpha \mathcal{L}' - \beta \mathcal{M}')Z \triangleq \alpha \begin{bmatrix} \searrow & \square \\ & \searrow \end{bmatrix} - \beta \begin{bmatrix} \searrow & \square \\ & \searrow \end{bmatrix}$$

and the upper left block of the pencil has only eigenvalues inside the unit circle.

Step 1:

Apply Algorithm 15.8 CSPHESP to compute $Q \in \mathcal{US}_{2n}(\mathbb{C})$ such that

$$Q^*(\alpha \mathcal{L}' - \beta \mathcal{M}')Q \triangleq \alpha \begin{bmatrix} \searrow \\ * & \searrow \end{bmatrix} - \beta \begin{bmatrix} \searrow & \square \\ & \searrow \end{bmatrix}$$

is a symplectic Hessenberg pencil.

Set $Z := Q$, $\mathcal{L}' := Q^* \mathcal{L}' Q$, $\mathcal{M}' = Q^* \mathcal{M}' Q$.

Step 2:

FOR $i = 1, 2, \ldots$

Set all subdiagonal elements of F satisfying

(15.31)
$$|f_{i,i-1}| \leq \epsilon(|f_{ii}| + |f_{i-1,i-1}|)$$

to zero. Find the largest $q \geq 0$ such that

(15.32)
$$\alpha \mathcal{L}' - \beta \mathcal{M}' = \alpha \begin{bmatrix} F_{11} & F_{12} & 0 & 0 \\ 0 & F_{22} & 0 & 0 \\ 0 & 0 & I & 0 \\ 0 & H_{22} & 0 & I \end{bmatrix} - \beta \begin{bmatrix} I & 0 & -G_{11} & -G_{12} \\ 0 & I & -G_{21} & -G_{22} \\ 0 & 0 & F_{11}^* & 0 \\ 0 & 0 & F_{12}^* & F_{22}^* \end{bmatrix} \begin{matrix} \}q \\ \}n-q \\ \}q \\ \}n-q \end{matrix}$$

such that F_{22} is unreduced upper Hessenberg.

IF $\quad q = n-1 \quad$ THEN triangularize

(15.33)
$$\alpha \begin{bmatrix} 1 & 0 & 0 & 0 \\ 0 & F_{22} & 0 & 0 \\ 0 & 0 & 1 & 0 \\ 0 & H_{22} & 0 & I \end{bmatrix} - \beta \begin{bmatrix} 1 & 0 & 0 & 0 \\ 0 & I & 0 & -G_{22} \\ 0 & 0 & 1 & 0 \\ 0 & 0 & 0 & F_{22}^* \end{bmatrix}$$

by applying a symplectic Givens congruence transformation with $Q_1 \in \mathcal{US}_{2n}(\mathbb{C})$ that eliminates H_{22}. Set $\mathcal{L}' = Q_1^* \mathcal{L}' Q_1$, $\mathcal{M}' = Q_1^* \mathcal{M}' Q_1$, $Z = ZQ_1$.
(We allow a "fill in" in block position (2.4) of $\mathcal{L}'$.)
Apply the standard QR-algorithm to F_{11} and compute $Q_2 \in \mathcal{U}_{n-1}(\mathbb{C})$ such that $Q_2^* F_{11} Q_2 \stackrel{\triangle}{=} \begin{bmatrix} \diagdown \end{bmatrix}$. Set

$$\tilde{Q}_2 = \mathrm{diag}(Q_2, 1, Q_2, 1), \ \mathcal{L}' := \tilde{Q}_2^* \mathcal{L}' \tilde{Q}_2, \ \mathcal{M}' = \tilde{Q}_2^* \mathcal{M}' \tilde{Q}_2, \ Z := Z\tilde{Q}_2.$$

QUIT

END IF

Set $\alpha \widetilde{\mathcal{L}}' - \beta \widetilde{\mathcal{M}}' = \alpha \begin{bmatrix} F_{22} & 0 \\ H_{22} & I \end{bmatrix} - \beta \begin{bmatrix} I & -G_{22} \\ 0 & F_{22}^* \end{bmatrix}$ and apply Algorithm CSP0DFP to $\alpha \widetilde{\mathcal{L}}' - \beta \widetilde{\mathcal{M}}'$ to compute S_1, Z_1 such that

$$S_1(\alpha \widetilde{\mathcal{L}}' - \beta \widetilde{\mathcal{M}}')Z_1 = \alpha \begin{bmatrix} \varphi_0 & \varphi_1^* & 0 & 0 \\ \varphi_2 & \tilde{F}_{22} & 0 & 0 \\ 0 & 0 & 1 & 0 \\ 0 & \tilde{H}_{22} & 0 & 1 \end{bmatrix} - \beta \begin{bmatrix} 1 & 0 & \gamma_1 & \gamma_2^* \\ 0 & 1 & \gamma_2 & -\tilde{G}_{22} \\ 0 & 0 & \varphi_0 & \varphi_2^* \\ 0 & 0 & \varphi_1 & \tilde{F}_{22}^* \end{bmatrix}.$$

IF $\quad \left\| \begin{bmatrix} \varphi_0 \\ \varphi_2 \end{bmatrix} \right\| < \epsilon \quad$ THEN set

$$\alpha \widehat{\mathcal{L}}' - \beta \widehat{\mathcal{M}}' := \begin{bmatrix} \tilde{F}_{22} & 0 \\ \tilde{H}_{22} & I \end{bmatrix} - \beta \begin{bmatrix} I & -\tilde{G}_{22} \\ 0 & \tilde{F}_{22}^* \end{bmatrix}$$

and $Z := Z \, \mathrm{diag}(I, Z_1, I, Z_1)$.
Compute $S_{11} = \tilde{F}_{12} + \tilde{G}_{22}\tilde{F}_{22}^{-*}\tilde{H}_{22}$, $S_{12} = \tilde{G}_{22}\tilde{F}_{22}^{-*}$ by forming a QR-decomposition $\tilde{F}_{22} = \tilde{Q}_{22}\tilde{R}_{22}$ and solving $S_{12}\tilde{R}_{22}^* = \tilde{G}_{22}\tilde{Q}_{22}$. Observe that $\tilde{G}_{22}\tilde{F}_{22}^{-*}\tilde{H}_{22} = \tilde{G}_{22}\tilde{Q}_{22}\tilde{R}_{22}^{-*}\tilde{H}_{22} = \tilde{\gamma}\tilde{G}_{22}\tilde{Q}_{22}e_ne_n^*$, where $\tilde{\gamma}$ is the last element of $\tilde{R}_{22}^{-*}\tilde{H}_{22}$.

END IF

END FOR

Step 3:

Apply Algorithm 15.25 CSPQR to

$$\widehat{S} = \begin{bmatrix} S_{11} & S_{12} \\ S_{21} & S_{22} \end{bmatrix},$$

where S_{21}, S_{22} are given implicitely via $\widetilde{H}_{22}, \widetilde{F}_{22}$ and the reduction step to symplectic Hessenberg form is omitted, and obtain Z_2 such that

$$Z_2^* \widehat{S} Z_2 \triangleq \begin{bmatrix} \widehat{T}_{11} & \widehat{T}_{12} \\ 0 & \widehat{T}_{11}^{-*} \end{bmatrix}$$

is in symplectic Schur-form and the matrix $\widehat{T}_{11}$ has only eigenvalues of modulus < 1.

Set $Z := Z \operatorname{diag}(I, 1, Z_2, I, 1, Z_2)$

END

Observe that except for the deflation of $0, \infty$ eigenvalues the right transformations are in $\mathcal{US}_{2n}(\mathbf{C})$, the only nonunitary matrix that occurs on the right is $\begin{bmatrix} I & H \\ 0 & I \end{bmatrix}$ which is well–conditioned. Clearly a corresponding algorithm for the real case, called SPQRP, is constructed easily in a similar way.

Algorithm 15.25 has the form discussed in Algorithm 14.14 except for the deflation step that deflates $0, \infty$ eigenvalues. So it has the required properties. It suffers, however, from the difficulty already discussed in Section 14, that roundoff destroys the symplectic structure and it is not easily detected what the deviation from the symplectic form is. A method that avoids this difficulty is currently not known but this topic is under investigation, see [F 2].

Another different approach to use unitary transformations to produce a Schur form is the Jacobi method, e.g. Golub/Van Loan [G 12]. A method has been proposed to adapt Jacobi's method for Hamiltonian matrices by Byers [B 42], in particular with the idea to use the great ability of Jacobi methods for parallel methods. Both methods still suffer from the general difficulties of applying Jacobi's method to unsymmetric matrices. This topic is still under investigation, thus we do not discuss it here.

We summarize the algorithms discussed in this Section in the following table, where we also give the operation counts. For the real case a flop is a real flop otherwise a complex flop. We always give two flop counts the upper one is for the updating of the matrix itself and the lower one for the accumulation of the transformation matrix.

15.34 TABLE. *Initial (Hessenberg) reduction algorithms*

Name of algorithm	function	input data	output data	flops
CHAMHES	complex reduction to Hamiltonian Hessenberg form	$\mathcal{H} = \begin{bmatrix} F & G \\ H & -F^* \end{bmatrix}$ $\in \mathbb{C}^{2n,2n}$ rank $H = 1$	$Q^*\mathcal{H}Q \hat{=} \begin{bmatrix} \boxed{} & \boxed{} \\ & \boxed{} \\ \ast & \end{bmatrix}$ $Q \in \mathcal{US}_{2n}(\mathbb{C})$	$\frac{8}{3}n^3$ $\frac{2}{3}n^3$
HAMHES	real reduction to Hamiltonian Hessenberg form	$\mathcal{H} = \begin{bmatrix} F & G \\ H & -F^T \end{bmatrix}$ $\in \mathbb{R}^{2n,2n}$	$Q^T\mathcal{H}Q \hat{=} \begin{bmatrix} \boxed{} & \boxed{} \\ & \boxed{} \\ \ast & \end{bmatrix}$ $Q \in \mathcal{US}_{2n}(\mathbb{R})$	$\frac{8}{3}n^3$ $\frac{2}{3}n^3$
CSPHESP	complex reduction to symplectic Hessenberg pencil	$\alpha \mathcal{L}' - \beta \mathcal{M}' =$ $\alpha \begin{bmatrix} F & 0 \\ H & I \end{bmatrix}$ $-\beta \begin{bmatrix} I & -G \\ 0 & F^* \end{bmatrix}$ $\in \mathbb{C}^{2n,2n}$ rank $H = 1$	$Q^*(\alpha\mathcal{L}' - \beta\mathcal{M}')Q \hat{=}$ $\alpha \begin{bmatrix} \boxed{} & \\ & \diagdown \\ \ast & \end{bmatrix} -$ $\beta \begin{bmatrix} \diagdown & \boxed{} \\ & \boxed{} \end{bmatrix}$ $Q \in \mathcal{US}_{2n}(\mathbb{C})$	$\frac{8}{3}n^3$ $\frac{2}{3}n^3$
SPHESP	real reduction to symplectic Hessenberg pencil	$\alpha \mathcal{L}' - \beta \mathcal{M}' =$ $\alpha \begin{bmatrix} F & 0 \\ H & I \end{bmatrix}$ $-\beta \begin{bmatrix} I & -G \\ 0 & F^* \end{bmatrix}$ $\in \mathbb{R}^{2n,2n}$ rank $H = 1$	$Q^T(\alpha\mathcal{L}' - \beta\mathcal{M}')Q \hat{=}$ $\alpha \begin{bmatrix} \boxed{} & \\ & \diagdown \\ \ast & \end{bmatrix} -$ $\beta \begin{bmatrix} \diagdown & \boxed{} \\ & \boxed{} \end{bmatrix}$ $Q \in \mathcal{US}_{2n}(\mathbb{R})$	$\frac{8}{3}n^3$ $\frac{2}{3}n^3$
CSPHES	complex reduction to symplectic Hessenberg matrix	$S = \begin{bmatrix} S_{11} & S_{12} \\ S_{21} & S_{22} \end{bmatrix}$ $\in \mathcal{S}_{2n}(\mathbb{C})$ rank $S_{21} = 1$	$Q^*SQ \hat{=} \begin{bmatrix} \boxed{} & \boxed{} \\ I & \boxed{} \end{bmatrix}$ $Q \in \mathcal{US}_{2n}(\mathbb{C})$	$\frac{14}{3}n^3$ $\frac{2}{3}n^3$

| SPHES | real reduction to symplectic Hessenberg matrix | $S = \begin{bmatrix} S_{11} & S_{12} \\ S_{21} & S_{22} \end{bmatrix} \in S_{2n}(\mathbf{C})$ rank $S_{21} = 1$ | $Q^T S Q \doteq \begin{bmatrix} \boxslash & \Box \\ | & \Box \end{bmatrix}$ $Q \in \mathcal{U}S_{2n}(\mathbf{R})$ | $\frac{14}{3} n^3$ $\frac{2}{3} n^3$ |

15.35 TABLE. *Implicit iteration steps*

Name of algorithm	function	input data/output data	forms	flops
CIMHQR	complex implicit Hamiltonian QR-step	$\mathcal{H} \in \mathbf{C}^{2n,2n}$ Hamiltonian Hessenberg $S \in \mathcal{U}S_{2n}(\mathbf{C})$	$\begin{bmatrix} \boxslash & \Box \\ * & \boxslash \end{bmatrix}$	$10\,n^2$ $30\,n^2$
IMHQR	real implicit Hamiltonian QR-step	$\mathcal{H} \in \mathbf{C}^{2n,2n}$ Hamiltonian Hessenberg $S \in \mathcal{U}S_{2n}(\mathbf{C})$	$\begin{bmatrix} \boxslash & \Box \\ * & \boxslash \end{bmatrix}$	$10\,n^2$ $30\,n^2$
CIMSPQR	complex implicit symplectic QR-step	$Y \in S_{2n}(\mathbf{C})$ in symplectic Hessenberg form $S \in \mathcal{U}S_{2n}(\mathbf{C})$	$\begin{bmatrix} \boxslash & \Box \\ | & \Box \end{bmatrix}$	$15\,n^2$ $30\,n^2$
IMSPQR	real implicit symplectic QR-step	$Y \in S_{2n}(\mathbf{R})$ in symplectic Hessenberg form $S \in \mathcal{U}S_{2n}(\mathbf{R})$	$\begin{bmatrix} \boxslash & \Box \\ | & \Box \end{bmatrix}$	$15\,n^2$ $30\,n^2$

15.36 TABLE. *Factorization algorithms*

Name	function	input	output
CHAMQR	complex Hamiltonian QR–algorithm	$\mathcal{H} \in \mathbf{C}^{n,n}$ Hamiltonian	Hamiltonian Schur–form of $\mathcal{H}$
HAMQR	real Hamiltonian QR–algorithm	$\mathcal{H} \in \mathbf{R}^{n,n}$ Hamiltonian	real Hamiltonian Schur–form of $\mathcal{H}$
CSPQR	complex symplectic QR–algorithm	$\mathcal{S} \in \mathcal{S}_{2n}(\mathbf{C})$	symplectic Schur–form of $\mathcal{S}$
SPQR	real symplectic QR–algorithm	$\mathcal{S} \in \mathcal{S}_{2n}(\mathbf{R})$	real symplectic Schur–form of $\mathcal{S}$
CSPQRP	complex symplectic QR–algorithm for pencils	$\alpha \mathcal{L}' - \beta \mathcal{M}'$ complex symplectic pencil	generalized symplectic Schur–form of $\alpha \mathcal{L}' - \beta \mathcal{M}'$
SPQRP	real symplectic QR–algorithm for pencils	$\alpha \mathcal{L}' - \beta \mathcal{M}'$ real symplectic pencil	generalized symplectic Schur–form of $\alpha \mathcal{L}' - \beta \mathcal{M}'$

§ 16 Nonunitary algorithms for real Hamiltonian or real symplectic eigenvalue problems.

In Algorithm 14.9 we have given a method that preserves the Hamiltonian or symplectic structure and is numerically strongly backwards stable. But in order to be practical this method needs an initial reduction to a Hessenberg like form and only in the special cases discussed in Section 15 such a reduction is known, and as was shown in Ammar/Mehrmann [A 4], there is not much hope to obtain such a reduction in a simple way. A possible way out of this dilemma is to relax the requirement that we only use unitary symplectic transformations.

It is generally possible to consider analogues to the QR- and QZ-algorithm, where the unitary matrices are replaced by symplectic matrices. This was first proposed by Della–Dora [D 1], [D 2] and shown to give a practical algorithm by Mehrmann [M 5]. Now the SR-decomposition and the transformation to Hessenberg form with symplectic matrices does not always exist, so precautions have to be taken to keep the algorithm as numerically stable as possible. The work and storage count gives a result that is comparable to that for the QR- and QZ-algorithm. Thus, for arbitrary matrices and pencils there is no reason to use these algorithms instead of the QR- or QZ-algorithm and we will therefore not discuss the general algorithm here. But for the Hamiltonian eigenproblem it has been shown by Bunse–Gerstner/Mehrmann [B 32], Bunse-Gerstner et al [B 36], that there exist very efficient variants of the SR-algorithm, so that the problem of not having unitary symplectic algorithms for general problems can be solved for the prize of working with nonunitary matrices in parts of the algorithms. For symplectic matrices and pencils, similar algorithms have been constructed in Flaschka et al [F 2]. A drawback of these type of algorithms is, that they are only applicable in the case of real Hamiltonian and symplectic matrices, which is, on the other hand, the case in many applications. We begin again with the Hamiltonian case. The SR-algorithm for Hamiltonian matrices was introduced by Bunse–Gerstner/Mehrmann [B 32]. It again consists of the two essential parts: Reduction to J-Hessenberg form and an implicit SR-step.

16.1 ALGORITHM JHESS [REDUCTION TO J–HESSENBERG FORM WITH OPTIMALLY CONDITIONED SYMPLECTIC TRANSFORMATIONS].

Given $\mathcal{H} = [h_{ij}] \in \mathbf{R}^{2n,2n}$, $S = I \in \mathbf{R}^{2n,2n}$ and a tolerance t for condition numbers of transformation matrices, the following algorithm overwrites $\mathcal{H}$ with $S_1 \mathcal{H} S_1^{-1} \stackrel{\triangle}{=}$

in J–Hessenberg form and S with $S\,S_1$, where $S_1 \in \mathcal{S}_{2n}(\mathbf{R})$. If in addition $\mathcal{H}$ is Hamiltonian then $S_1 \mathcal{H} S_1^{-1} \stackrel{\triangle}{=}$.

(*) FOR $\quad i = 1, \ldots, n-1$

Let $Z = \begin{bmatrix} x \\ y \end{bmatrix} = \mathcal{H} e_i$ and set $Z_i = HHGEN(i+1,n,y)$.

Set $\mathcal{H} := \begin{bmatrix} Z_i^T & 0 \\ 0 & Z_i^T \end{bmatrix} \mathcal{H} \begin{bmatrix} Z_i & 0 \\ 0 & Z_i \end{bmatrix}$, $\quad S := S \begin{bmatrix} Z_i & 0 \\ 0 & Z_i \end{bmatrix}$.

Let $z = \mathcal{H}e_i$ and $Z = GSGEN(i+1, z)$.

Set $\mathcal{H} = Z^*\mathcal{H}Z$, $S := SZ$.

Let $\begin{bmatrix} x \\ y \end{bmatrix} = \mathcal{H}e_i$ and $Q_i = HHGEN(i+1, n, x)$.

Set $\mathcal{H} := \begin{bmatrix} Q_i^T & 0 \\ 0 & Q_i^T \end{bmatrix} \mathcal{H} \begin{bmatrix} Q_i^T & 0 \\ 0 & Q_i \end{bmatrix}, S := S \begin{bmatrix} Q_i & 0 \\ 0 & Q_i \end{bmatrix}$.

IF $(h_{i+1,i} \neq 0$ AND $h_{n+i,i} = 0)$ THEN

    GOTO (**)

ELSE

    Let $z = \mathcal{H}e_i$ and $G_i = G0GEND(i+1, z)$.

    IF   cond $_2(G_i) > t$ THEN

        GOTO (**)

    END IF

END IF

Set $\mathcal{H} := G_i\mathcal{H}G_i^{-1}$, $S := SG_i^{-1}$.

Let $\begin{bmatrix} x \\ y \end{bmatrix} = \mathcal{H}e_{n+i}$ and $Z_i = HHGEN(i+1, n, y)$.

Set $\mathcal{H} := \begin{bmatrix} Z_i^T & 0 \\ 0 & Z_i^T \end{bmatrix} \mathcal{H} \begin{bmatrix} Z_i & 0 \\ 0 & Z_i \end{bmatrix}, S := S \begin{bmatrix} Z_i & 0 \\ 0 & Z_i \end{bmatrix}$.

Let $z = \mathcal{H}e_{n+i}$ and $Z = GSGEN(i+1, z)$.

Set $\mathcal{H} := Z^T\mathcal{H}Z$, $S := SZ$.

IF $i \leq n-2$ THEN

    Let $\begin{bmatrix} x \\ y \end{bmatrix} = \mathcal{H}e_{n+i}$ and $Q_i = HHGEN(i+1, n, x)$.

    Set $\mathcal{H} := \begin{bmatrix} Q_i^T & 0 \\ 0 & Q_i^T \end{bmatrix} \mathcal{H} \begin{bmatrix} Q_i & 0 \\ 0 & Q_i \end{bmatrix}, S := S \begin{bmatrix} Q_i & 0 \\ 0 & Q_i \end{bmatrix}$.

END IF

END FOR

(**) Set $\mathcal{H} := S_k\mathcal{H}S_k^{-1}$, $S := SS_k^{-1}$ for some random $S_k \in \mathcal{S}_{2n}(\mathbf{R})$.

GOTO (*)

END

Stability problems in this algorithm can only occur, when transforming with a matrix G_i, that has a large condition number or if G_i does not exist. To monitor this problem, we choose a certain tolerance for the condition number of the nonunitary transformations. If the condition number exceeds this bound at some stage, then we choose a random symplectic transformation and start the procedure anew. For a justification that this is working and does not lead to an infinite process, see Bunse–Gerstner [B 23] or Bunse–Gerstner et al [B 36]. One step of the iterative part is then as follows:

16.2 ALGORITHM IMOCSR [IMPLICIT OPTIMALLY CONDITIONED SR STEP FOR REAL HAMILTONIAN MATRICES].

Given a real Hamiltonian matrix $\mathcal{H}$ in J–Hessenberg form, and given $\lambda_1 \in \mathbb{C}\backslash\{i\alpha|\alpha \in \mathbb{R}\}$ or $\lambda_1, \lambda_2 \in \mathbb{R}$ approximations to eigenvalues of $\mathcal{H}$, $S \in \mathcal{S}_{2n}(\mathbb{R})$ and a tolerance for condition numbers t, the following algorithm overwrites $\mathcal{H}$ with $S_1^{-1}\mathcal{H}S_1$ and S with SS_1, where $S_1 \in \mathcal{S}_{2n}(\mathbb{R})$ is a product of Householder symplectic, Givens symplectic matrices and elementary symplectic matrices of type G_0 and $S_1^{-1}p(\mathcal{H}) = R_1$, where R_1 is J–triangular and $p(\mathcal{H})$ is a polynomial in $\mathcal{H}$. Usually

(16.3) $$p(\mathcal{H}) = (\mathcal{H} - \lambda_1 I)(\mathcal{H} + \lambda_2 I) \text{ for } \lambda_1, \lambda_2 \in \mathbb{R}$$

or

(16.4) $$p(\mathcal{H}) = (\mathcal{H} + \lambda_1 I)(\mathcal{H} - \lambda_1 I)(\mathcal{H} + \overline{\lambda}_1 I)(\mathcal{H} - \overline{\lambda}_1 I) \text{ for } \lambda_1 \neq \alpha, i\alpha \text{ for } \alpha \in \mathbb{R}.$$

(*) Let $Z = p(\mathcal{H})e_1 = \begin{bmatrix} x \\ y \end{bmatrix}$.

(Due to the special form of $\mathcal{H}$ it follows that $y = 0$ and we have $x = [x_1, \ldots, x_r, 0, \ldots, 0]^T$, where $r = 2$ if $p(\mathcal{H})$ is as in (16.3) or $r = 3$ if $p(\mathcal{H})$ is as in (16.4).)

Let $Z = HHGEN(1, r, x)$ and set $\mathcal{H} := \begin{bmatrix} Z^T & 0 \\ 0 & Z^T \end{bmatrix} \mathcal{H} \begin{bmatrix} Z & 0 \\ 0 & Z \end{bmatrix}$, $S := S \begin{bmatrix} Z & 0 \\ 0 & Z \end{bmatrix}$.

(We then reduce $\mathcal{H}$ again to J–Hessenberg form, using the following simplified version of Algorithm 16.1 JHESS.)

FOR $i = 1, \ldots, n-1$

    Let $m = \min(i + r, n)$.

    FOR $k = m, \ldots, i+1$

        Let $z = \mathcal{H}e_i$ and let $Z_k = GSGEN(k, z)$.

        Set $\mathcal{H} := Z_k^T \mathcal{H} Z_k$, $S := SZ_k$.

    END FOR

    Let $\begin{bmatrix} x \\ y \end{bmatrix} = \mathcal{H}e_i$ and $Q_i = HHGEN(i+2, i+r, x)$.

    Set $\mathcal{H} := \begin{bmatrix} Q_i^T & 0 \\ 0 & Q_i^T \end{bmatrix} \mathcal{H} \begin{bmatrix} Q_i & 0 \\ 0 & Q_i \end{bmatrix}$, $S := S \begin{bmatrix} Q_i & 0 \\ 0 & Q_i \end{bmatrix}$.

    IF $h_{i+1,i} \neq 0$ AND $h_{n+i,n+1} = 0$ THEN

        Choose different $\lambda_1, (\lambda_2)$

        GOTO (*)

ELSE

    Let $z = \mathcal{H}e_i$ and $G_i = G0GEND(i+1, z)$.

    IF   cond $_2(G_i) > t$ THEN

        choose different $\lambda_1(\lambda_2)$

        GOTO (*)

    END IF

    Set $\mathcal{H} := G_i \mathcal{H} G_i^{-1}$, $S := S G_i^{-1}$.

END IF

FOR $k = m, \ldots, i+1$

    Let $z = \mathcal{H}e_{n+i}$ and $Z_k = GSGEN(k, z)$.

    Set $\mathcal{H} := Z_k^T \mathcal{H} Z_k$, $S := S Z_k$.

END FOR

IF $i \leq n - 2$ THEN

    Let $\begin{bmatrix} x \\ y \end{bmatrix} = \mathcal{H}e_{n+i}$ and $Q_i = HHGEN(i+1, n, x)$.

    Set $\mathcal{H} := \begin{bmatrix} Q_i^T & 0 \\ 0 & Q_i^T \end{bmatrix} \mathcal{H} \begin{bmatrix} Q_i & 0 \\ 0 & Q_i \end{bmatrix}$, $S := S \begin{bmatrix} Q_i & 0 \\ 0 & Q_i \end{bmatrix}$.

END IF

END FOR

END

The SR-algorithm for Hamiltonian matrices is then:

16.5 ALGORITHM HAMSR [HAMILTONIAN SR ALGORITHM].

Given $\mathcal{H} \in \mathbf{R}^{2n,2n}$ Hamiltonian, having no eigenvalues on the imaginary axis, a tolerance $\epsilon > 0$ and a tolerance t for the condition number of transformation matrices . Let $S = I \in \mathbf{R}^{2n,2n}$. The following algorithm overwrites $\mathcal{H}$ with $Z^{-1}\mathcal{H}Z = \begin{bmatrix} D_1 & D_2 \\ 0 & -D_1^T \end{bmatrix}$, where D_1, D_2 are block diagonal with blocks of size 1×1 or 2×2, and S with $SZ \in S_{2n}(\mathbf{R})$.

Step 1: Transform $\mathcal{H}$ to J–triangular form using Algorithm JHESS.

Step 2:

    FOR $j = 1, 2 \ldots$

        Let $\mathcal{H} = \begin{bmatrix} F & G \\ H & -F^T \end{bmatrix}$. Set all subdiagonal of G satisfying

$$|g_{i,i-1}| < \epsilon(|g_{i,i}| + |g_{i-1,-1}|)$$

    to zero. Do the same with $g_{i-1,i}$.

Let $q \geq 0$ be the largest integer and $p \geq 0$ the smallest integer such that

$$
\mathcal{H} = \begin{bmatrix} F_{11} & & & G_{11} & & \\ & F_{22} & & & G_{22} & \\ & & F_{33} & & & G_{33} \\ H_{11} & & & -F_{11}^T & & \\ & H_{22} & & & -F_{22}^T & \\ & & H_{33} & & & -F_{33}^T \end{bmatrix} \begin{matrix} \}q \\ \}p \\ \}n-p-q \\ \}q \\ \}p \\ \}n-p-q \end{matrix} ,
$$

where G_{33} is block diagonal with blocks of size ≤ 2 and G_{22} is unreduced.

IF $q = n$ THEN

     Compute S_T such that $\mathcal{H} := S_T^{-1} \mathcal{H} S_T = \begin{bmatrix} T_{11} & T_{12} \\ 0 & -T_{11}^T \end{bmatrix}$, $S := S S_T$ and T_{11} has only eigenvalues with negative real part.

     QUIT

ELSE

     Apply Algorithm 16.2 IMHAMSR to $\begin{bmatrix} F_{22} & G_{22} \\ H_{22} & -F_{22}^T \end{bmatrix}$ and compute

$$
S_1 = \begin{bmatrix} S_{11} & S_{12} \\ S_{21} & S_{22} \end{bmatrix} \in \mathcal{S}_{2(n-p-q)}(\mathbf{R}), \quad \tilde{S}_1 = \begin{bmatrix} I & 0 & 0 & 0 & 0 & 0 \\ 0 & S_{11} & 0 & 0 & S_{12} & 0 \\ 0 & 0 & I & 0 & 0 & 0 \\ 0 & 0 & 0 & I & 0 & 0 \\ 0 & S_{21} & 0 & 0 & S_{22} & 0 \\ 0 & 0 & 0 & 0 & 0 & I \end{bmatrix}.
$$

     Set $\mathcal{H} := \tilde{S}_1^{-1} \mathcal{H} \tilde{S}_1$, $S := S \tilde{S}_1$.

     END IF

   END FOR

END

For a detailed description of the part where $\mathcal{H}$ is transformed to $\begin{bmatrix} T_{11} & T_{12} \\ 0 & -T_{11}^T \end{bmatrix}$ and any other details see Bunse–Gerstner/Mehrmann [B 32].

The main objection against this algorithm is the fact that nonorthogonal transformations are used, that may cause huge roundoff errors due to ill–conditioning of the transformations. The exceptional shift techniques described in the algorithm are a way out of this dilemma, which is quite satisfactory, but the numerical experience shows that the results are often less accurate than those produced by the QR-algorithm of Laub [L 7].

Thus, when this method is used for the computation of deflating subspaces or solutions to Riccati equations, it is advisable to refine the solution with a few steps of defect correction. If one is anyway refining or reiterating, one might as well drop further restrictions on the transformation matrices, i.e. work with Gauss type transformations all the way, i.e. replace the orthogonal transformations by lower and upper triangular transformations to decrease the number of operations. This was proposed in connection with other simplifications by Bunse–Gerstner et al [B 36] and we omit this simplification here.

For the SR-algorithm we have not yet discussed the choice of shifts. Keeping in mind that the J–Hessenberg form is just a permutation of the usual Hessenberg form, (see Section 7), and the shifting is done essentially as in the QR-algorithm, one obvious choice would be the analogoue to the Wilkinson shift, which is here as follows. For a J–tridiagonal matrix

$$\mathcal{H} = \begin{bmatrix} A & G \\ H & -A^T \end{bmatrix}$$

with A, H diagonal, G a symmetric tridiagonal, consider the 4×4 submatrix consisting of the 2×2 lower right corner matrices of the four blocks, which is again Hamiltonian and compute its eigenvalues (see Bunse–Gerstner/Mehrmann [B 32]). If they are nonreal and not purely imaginary, then they are $\lambda, \bar{\lambda}, -\lambda, -\bar{\lambda}$, otherwise we have $\lambda_1, -\lambda_1, \lambda_2, -\lambda_2$. Depending on the reality of λ_1, λ_2, we choose the different shift polynomials in (16.3), (16.4).

We now discuss the symplectic case. Observe that the algorithms JHESS, GJHESS are applicable for general matrices in $\mathbf{R}^{2n,2n}$, so if $S \in S_{2n}(\mathbf{R})$ they yield the J–Hessenberg

form $S \triangleq \begin{bmatrix} \diagdown & \diagdown \\ \diagdown & \diagdown \end{bmatrix}$.

For symplectic matrices now, no further zeros occur in S, that simplify the algorithm but applying Proposition 15.10, we see that S actually only depends on $4n - 1$ parameters.

16.7 PROPOSITION. Let $S = \begin{bmatrix} S_{11} & S_{12} \\ S_{21} & S_{22} \end{bmatrix} \in S_{2n}(\mathbf{R})$ be in J–Hessenberg form and S_{22} nonsingular then

(16.8) $\qquad G = [g_{ij}] = S_{12}S_{22}^{-1}$ is symmetric tridiagonal ,

(16.9) $\qquad H = [h_{ij}] = S_{22}^{-1}S_{21}$ is diagonal ,

(16.10) $\qquad F = [F_{ij}] = S_{22}^{-T}$ is bidiagonal, and

(16.11) $\qquad f_{i,i-1} = -g_{i,i-1}h_{i-1,i-1}/f_{i-1,i-1}, \ i = 2, \dots, n.$

PROOF: Let

$$S = \begin{bmatrix} I & G \\ 0 & I \end{bmatrix} \begin{bmatrix} F & 0 \\ 0 & F^{-T} \end{bmatrix} \begin{bmatrix} I & 0 \\ H & I \end{bmatrix}$$

where $F = S_{22}^{-T}$, $G = S_{12}S_{22}^{-1}$, $H = S_{22}^{-1}S_{21}$ be the decomposition of Proposition 15.10 of S for invertible S_{22}. Now $S_{22} \triangleq \begin{bmatrix} \diagdown \end{bmatrix}$ and hence $F^T = S_{22}^{-1} \triangleq \begin{bmatrix} \diagdown \end{bmatrix}$. $G =$

$S_{12}S_{22}^{-1} \triangleq$  is symmetric and thus tridiagonal. $H = S_{22}^{-1}S_{21} \triangleq$  is symmetric and thus diagonal, and since $S_{21} \triangleq$  we have

$$S_{11} = F + GF^{-T}H \triangleq$$ 

It follows that $F^T \triangleq$  and F has as subdiagonal the negative subdiagonal of $GF^{-T}H$, thus

$$f_{i,i-1} = -g_{i,i-1}h_{i-1,i-1}/f_{i-1,i-1}, \quad i = 2, \ldots, n.$$

■

An algorithm that employs this parametrization is discussed and analyzed in Flaschka et al [F 2]. If we use the parametrization of Proposition 16.7 and compute consecutive steps of the SR–algorithm by only updating these parameters, we cannot guarantee that this updating always exists, see Flaschka et al [F 2]. Thus, apart from the problem of nonexisting transformations as in the Hamiltonian case, which can be treated by exceptional shifts, one has the problem that in order to determine the next set of parameters one has to compute $2n - 1$ decomposition of the type discussed in Proposition 15.10.

Flaschka et al [F 2] draw the conclusion that in the symplectic case such a method is numerically not acceptable, even if it is combined with defect correction.

Another method that uses nonunitary transformations is the HHDR algorithm of Bunse–Gerstner/Mehrmann [B 31] for general Hermitian pencils, which may be applied to the continuous time problem. Unfortunately this transformation turns real Hamiltonian into complex indefinite Hermitian pencils. Thus, only if the problem was nonreal in the beginning it is as efficient as QZ-algorithm. We do not discuss this procedure here, see Bunse–Gerstner/Mehrmann [B 31].

Another possibility to reduce the amount of work per iteration step of a subspace method like the Hamiltonian QR- or SR-algorithm is to replace the accumulation of the transformation matrices by the following direct updating method proposed by Byers/Mehrmann [B 45], which can be applied efficiently for Hamiltonian matrices, symplectic pencils or matrices. We have seen in Section 7, that if $S = \begin{bmatrix} S_{11} & S_{12} \\ S_{21} & S_{22} \end{bmatrix}$ is the right transformation matrix in the transformation to Schur–form, then the Hermitian matrix

$$X = -S_{21}\, S_{11}^{-1} \tag{16.12}$$

is the solution of the corresponding algebraic Riccati equation. Now observe that if $S \in \mathcal{S}_{2n}(\mathbb{C})$ and S_{11}^{-1} exists, then always $-S_{21}\, S_{11}^{-1}$ is Hermitian. But our transformation matrices are in $\mathcal{S}_{2n}(\mathbb{C})$ at any stage of the algorithm. This leads to the idea to update $X = -S_{21}\, S_{11}^{-1}$ at every stage of the algorithm instead of first accumulating all the transformation matrices and then computing the matrix X. To do this, one only has to introduce formulas for the updating with the elementary symplectic transformations introduced in Section 13. Following Byers/Mehrmann [B 45] we set

$$S := \begin{bmatrix} S_{11} & S_{12} \\ S_{21} & S_{22} \end{bmatrix}, \; X := -S_{21}\, S_{11}^{-1}, \; Y = S_{11}^{-1}\, S_{12}, \; Z = S_{11}^{-1}.$$

Initially we have $X = 0, Y = 0, Z = I$. If T is the matrix to update with, then set

$$\widetilde{S} = \begin{bmatrix} \widetilde{S}_{11} & \widetilde{S}_{12} \\ \widetilde{S}_{21} & \widetilde{S}_{22} \end{bmatrix} = ST, \ \widetilde{X} = -\widetilde{S}_{21}\widetilde{S}_{11}^{-1}, \ \widetilde{Y} = \widetilde{S}_{11}^{-1}\widetilde{S}_{12}, \ \widetilde{Z} = \widetilde{S}_{11}^{-1}.$$

We then have the following formulae:

a) Updating with Householder symplectic matrices $T = \begin{bmatrix} Q & 0 \\ 0 & Q \end{bmatrix}$, $Q \in U_n(\mathbf{C})$.

Then,

(16.13) $$\widetilde{Z} = QZ, \ \widetilde{X} = X, \ \widetilde{Y} = QYQ^*.$$

b) Updating with Givens symplectic matrices $T = \begin{bmatrix} C & S \\ -S & C \end{bmatrix}$, $C = I + (c-1)\, e_k e_k^T$, $S = -s\, e_k e_k^T$, $|c|^2 + |s|^2 = 1$, $\bar{c}s \in \mathbf{R}$.

Then,

$$\widetilde{Z} = (I - \frac{1}{\alpha}(c-1)\, e_k e_k^* - \frac{c}{s}\alpha\, Y e_k e_k^*)Z,$$

(16.14) $$\widetilde{X} = X - \frac{s}{\alpha}\, Z^* e_k e_k^* Z$$

$$\widetilde{Y} = Y + \frac{1}{\alpha}[e_k e_k^*(-s + 2(c-1)y_{kk}) + (1-c)(Y e_k e_k^* + e_k e_k^* Y) + s\, Y e_k e_k^* Y],$$

where $\alpha = c + s y_{kk}$.

c) Updating with an optimally conditioned matrix of type G_0, $T = \begin{bmatrix} D & DV_0 \\ 0 & D^{-1} \end{bmatrix}$.

Then,

(16.15) $$\widetilde{Z} = DZ, \ \widetilde{X} = X, \ \widetilde{Y} = -DV_0 D + DYD.$$

Clearly analogous formulae hold in the real case. The only critical updating formula is (16.14) if $|\alpha|$ is small. If such a small $|\alpha|$ occurs, then this updating formula is numerically unstable. One can again avoid such a failure by exceptional transformations. For details see Byers/Mehrmann [B 45] and Bunse–Gerstner/Mehrmann [B 32]. Observe that X, Y are Hermitian and thus only half of the matrix has to be computed and stored. The updating procedures are easily implemented using the given formulas. If we replace accumulation and subsequent computation of X (denoted by XSOLVE) by this updating procedure, we denote this by UPDAT.

Another interesting feature of these formulas is that the cost intensive parts in a), b), consist only of rank 1 modifications, which can be implemented easily on vector or parallel computers. We will not go into detail on this topic here, but for large systems, this may decrease the computing time of the most expensive part of the method. This is a topic for future investigations.

In this section we have discussed nonunitary algorithms for the computation or invariant subspaces of Hamiltonian matrices or deflating subspaces of symplectic pencils. The Hamiltonian SR-algorithm is a relatively cheap method that, combined with defect correction, yields satisfactory results. The cost can be reduced as in the version introduced by Bunse–Gerstner et al [B 36]. In the symplectic case, the corresponding method is less attractive as the analysis of Flaschka et al [F 2] shows. We finish this Section with a table of algorithms analogous to that in Section 15.

16.16 TABLE.

Name of algorithm	function	input data	output data	flops
JHESS	Reduction to J–tridiagonal form with optimally conditioned transformations	$\mathcal{H} \in \mathbf{R}^{2n,2n}$ Hamiltonian	$S^{-1}\mathcal{H}S \overset{\Delta}{=}$ $\left[\begin{smallmatrix}\ \end{smallmatrix}\right]$ $S \in \mathcal{S}_{2n}(\mathbf{R})$	$\frac{20}{3}\,n^3$ $12\,n^3$
IMHAMSR	implicit SR step with optimally conditioned transformations	$\mathcal{H} \in \mathbf{R}^{2n,2n}$	$S^{-1}\mathcal{H}S \overset{\Delta}{=}$ $\left[\begin{smallmatrix}\ \end{smallmatrix}\right]$ J–triangular $S \in \mathcal{S}_{2n}(\mathbf{R})$	$O(n)$ $24\,n^2$

§ 17 Closed loop algorithms.

We want to compute invariant subspaces of Hamiltonian or symplectic matrices and deflating subspaces of symplectic pencils. Thus, we are not primiarily interested in the eigenvalues
themselves. These eigenvalues will be the eigenvalues of the closed loop system matrix of
the controlled system. Their location therefore determines the asymptotic behaviour of
the controlled system. But they are also the best choices of shifts for the factorization
algorithms discussed in Sections 14–16, since in theory, with these shifts, the discussed
algorithms converge in at most n steps, and with these shifts an ordering at the end of the
procedure can be avoided. On the other hand they are a by product of all the algorithms
for the computation of Schur–like forms that we have discussed. But for Hamiltonian
and symplectic eigenvalue problems we have the discussed difficulties to obtain a practical
numerically strongly backwards stable method. There exist, however, reliable numerical
methods for the computation of the closed loop eigenvalues due to Van Loan [V 7] in the
real Hamiltonian case and Lin [L 21] in the real symplectic case. For the Hamiltonian case,
the square reduced algorithm of Van Loan [V 7] is based on the following observation:

Let $\mathcal{H}^s = \{ W \in \mathbf{R}^{2n,2n} | W = \mathcal{H}^2$ for some Hamiltonian matrix $\mathcal{H} \}$. If $W = \mathcal{H}^2 \in \mathcal{H}^s$
and $S \in S_{2n}(\mathbf{R})$ then $S^{-1} W S \in \mathcal{H}^s$ and if $\lambda_1, \dots, \lambda_n, -\lambda_1, \dots, -\lambda_n$ are the eigenvalues
of $\mathcal{H}$, then $\lambda_1^2, \lambda_1^2, \dots, \lambda_n^2, \lambda_n^2$ are the eigenvalues of W.

Thus, similarity transformations with symplectic matrices also leave $\mathcal{H}^s$ invariant. But if

$$\mathcal{H} = \begin{bmatrix} F & G \\ H & -F^T \end{bmatrix} \in \mathbf{R}^{2n,2n}$$

is Hamiltonian then in

$$(17.1) \qquad \mathcal{H}^2 = \begin{bmatrix} F^2 + GH & GF - GF^T \\ HF - F^T H & HG + (F^T)^2 \end{bmatrix}$$

the (2,1) block is skew symmetric. If we now apply the Hessenberg like reduction of
Paige/Van Loan [P 2] given by Algorithm 14.17 HLRED (CHLRED) to the square of a
real Hamiltonian matrix as in (17.1) then since the (2,1) block is real skew symmetric and
stays real skew symmetric, it annihilates the (2,1) block in (17.1). This is the idea of the
square reduced algorithm of Van Loan [V 7].

17.2 ALGORITHM SQRED [SQUARE REDUCED ALGORITHM FOR THE COMPUTATION OF
THE EIGENVALUES OF A REAL HAMILTONIAN MATRIX].

Given a Hamiltonian matrix $\mathcal{H} = \begin{bmatrix} F & G \\ H & -F^T \end{bmatrix} \in \mathbf{R}^{2n,2n}$ this algorithms computes the
eigenvalues of $\mathcal{H}$.

Step 1: Form
$$P = \mathcal{H}^2 = \begin{bmatrix} F^2 + GH & FG - GF^T \\ HF - F^T H & HG + (F^T)^2 \end{bmatrix}.$$

Step 2: Apply algorithm 14.17 HLRED to P, and determine $Q \in US_{2n}(\mathbf{R})$ such that

$$Q^* P Q = \begin{bmatrix} F_1 & G_1 \\ 0 & -F_1^T \end{bmatrix}.$$

Step 3:

Apply the QR-algorithm to the matrix F_1 (acumulation of the transformation matrices not required) to compute the eigenvalues $\mu_1, \ldots, \mu_n$ of F_1.

Step 4:

Set $\lambda_i = \sqrt{\mu_i}$ taking the square root in the left half plane and set $\lambda_{i+n} = -\lambda_i$ for $i = 1, \ldots, n$.

END

Van Loan [V 7] also proposes an implicit version of this procedure, which uses less storage and costs approximately the same. The computational cost of this method is $20n^2$ flops.

Steps 2, 3 in Algorithm 17.2 are numerically strongly backwards stable, the critical part in this algorithm is Step 1. As is well-known, e.g. Golub/Van Loan [G 12], roundoff errors in squaring a matrix can introduce large errors, so it is possible that the resulting eigenvalues may be less accurate than a numerically stable method would yield for the matrix itself.

Another approach to compute the eigenvalues for the use as 'exact' shifts is to run the SR-algorithm HAMSR without accumulating the transformations. This is cheaper than Van Loan's algorithm but less stable. A third method would clearly be the use of the standard QR-algorithm without accumulation of the transformation matrices, thereby ignoring the Hamiltonian structure. This approach can lead to difficulties in deciding which are the unstable eigenvalues. In Van Loan [V 7] an example is mentioned, where the standard QR-algorithm produced $2n - 1$ eigenvalues in the left half plane and one in the right half plane.

If we wish to use these eigenvalues as shifts to speed up convergence or to compute the deflating subspace without going through the whole iteration, then a loss in accuracy will not be so drastic, in particular since we have a good iterative refinement method at hand. (See Section 10).

For the discrete case an algorithm to compute the eigenvalues of a real symplectic pencil

$$(17.3) \qquad \mathcal{L}' - \lambda \mathcal{M}' =: \begin{bmatrix} F & 0 \\ H & I \end{bmatrix} - \lambda \begin{bmatrix} I & -G \\ 0 & F^T \end{bmatrix}$$

was proposed by Lin [L 21]. The key idea of this method is to compute instead of the eigenvalues $\lambda_1, \ldots, \lambda_{2n}$ of $\mathcal{L}' - \lambda \mathcal{M}'$ the eigenvalues $\mu_1, \ldots, \mu_{2n}$ of

$$(17.4) \qquad P - \mu Q = J^T[(\mathcal{L}' J (\mathcal{M}')^T + \mathcal{M}' J (\mathcal{L}')^T) - \mu \mathcal{M}' J (\mathcal{M}')^T].$$

Observe that if λ_j is an eigenvalue of $\mathcal{L}' - \lambda \mathcal{M}'$, then $\mu_j = \lambda_j + \frac{1}{\lambda_j}$ is an eigenvalue of $P - \mu Q$. Now

$$(17.5) \qquad P - \mu Q = \begin{bmatrix} GF^T - FG & F^2 + GH + I \\ -(I + HG + (F^T)^2) & HF - F^T H \end{bmatrix} - \mu \begin{bmatrix} 0 & F \\ -F^T & 0 \end{bmatrix}.$$

Set $W = F^2 + GH + I$, $Y = FG - GF^T$, $Z = HF - F^T H$. Then $P - \mu Q$ is equivalent to

$$(17.6) \qquad \widehat{P} - \mu \widehat{Q} = \begin{bmatrix} W^T & Z \\ Y & W \end{bmatrix} - \mu \begin{bmatrix} F^T & 0 \\ 0 & F \end{bmatrix}$$

where Y, Z are skew symmetric.

Observe that F may still be singular.

Using the deflation procedure of Algorithm 15.17 CSPDEF, we can deflate all the eigenvalues $0, \infty$ of this pencil (17.4) in advance and can then assume that F is nonsingular. In the algorithm proposed by Lin [L 21] this deflation is avoided and F is diagonalized by the QR-decomposition with column pivoting followed by backward substitution, e.g. Golub/Van Loan [G 12], which transforms $\widehat{P} - \mu \widehat{Q}$ to the form

$$(17.7) \qquad \widetilde{P} - \mu \widetilde{Q} = \begin{bmatrix} \widetilde{A}^T & \widetilde{C} \\ \widetilde{B} & \widetilde{A} \end{bmatrix} - \mu \begin{bmatrix} \widetilde{F}^T & 0 \\ 0 & \widetilde{F} \end{bmatrix}$$

with $\widetilde{F}$ diagonal. Then a variation of the MDR–algorithm of Bunse–Gerstner [B 24] is used to bring $\widetilde{A}$ to lower Hessenberg form and to eliminate $\widetilde{B}$ while keeping $\widetilde{F}$ diagonal. Observe that

$$(17.8) \quad \widetilde{\mathcal{H}} - \mu I = \begin{bmatrix} \widetilde{F}^{-T} & 0 \\ 0 & I \end{bmatrix} (\widetilde{P} - \mu \widetilde{Q}) \begin{bmatrix} I & 0 \\ 0 & \widetilde{F}^{-1} \end{bmatrix} = \begin{bmatrix} \widetilde{F}^{-T} \widetilde{A}^T & \widetilde{F}^{-T} \widetilde{C} \widetilde{F}^{-1} \\ \widetilde{B} & \widetilde{A} \widetilde{F}^{-1} \end{bmatrix} - \mu I$$

and $\mathcal{H}$ is skew Hamiltonian, i.e. $\mathcal{H} J = -(\mathcal{H} J)^T$.

Applying Algorithm 14.7 HLRED to $\mathcal{H}$ we get

$$\widehat{\mathcal{H}} = \begin{bmatrix} \widehat{A} & \widehat{C} \\ 0 & \widehat{A}^T \end{bmatrix} \;\; \widehat{=} \;\; \begin{bmatrix} \diagdown & \square \\ 0 & \diagdown \end{bmatrix}$$

and thus, we may apply steps 2, 3 of Algorithm 17.3 SQRED to $\widehat{\mathcal{H}}$ too, to compute the eigenvalues. Lin's algorithm, which we denote by LINRED has the advantage that it avoids the inversion of $\widetilde{F}$ and uses transformations which are bounded in norm, it does not avoid the inaccuracies already occuring when forming (17.7), which is the analogue to forming the square in the Hamiltonian case. Lin's algorithm costs approximately $20n^3$ flops. For further details on this method see Lin [L 21]. Again we may also use the standard QZ-algorithm without accumulation of transformation matrices to compute the closed loop eigenvalues, thereby ignoring the symplectic structure. We then may have again the problem to decide which are the stable and unstable eigenvalues.

In this section we have discussed methods to compute the eigenvalues of Hamiltonian matrices or symplectic pencils with cost effective numerically realiable methods. The computed eigenvalues can then serve as shifts for other algorithms like the one we discuss in the next section.

§ 18 A combination algorithm for real Hamiltonian and symplectic eigenvalue problems.

We have seen in Section 14 that the missing ingredient for a practical and reliable numerical method to solve the Hamiltonian and symplectic eigenvalue problem is an initial reduction to a Hessenberg like form that stays invariant under the iteration of Algorithm 14.9. We have seen that the Hamiltonian Hessenberg form and symplectic Hessenberg forms are the ideal candidates for such Hessenberg like forms.

But in Ammar/Mehrmann [A 4] it was shown that the existence of such a transformation is related to the solution of a set of nonlinear equations.

18.1 THEOREM.

i) Let $\mathcal{H} \in \mathbf{R}^{2n,2n}$ be Hamiltonian. Then there exists an orthogonal and symplectic matrix $Q \in \mathbf{R}^{2n,2n}$ such that $Q^T \mathcal{H} Q$ is an unreduced Hamiltonian Hessenberg matrix **if and only if** the nonlinear system of equations

$$(18.2) \qquad \begin{aligned} x^T J \mathcal{H}^{2i-1} x &= 0, \quad i = 1, \ldots, n-1 \\ x^T x &= 1 \end{aligned}$$

has a solution vector that is not contained in an invariant subspace of $\mathcal{H}$ of dimension less than or equal to n.

ii) Let $S \in \mathbf{R}^{2n,2n}$ be symplectic. Then there exists an orthogonal and symplectic matrix $Q \in \mathbf{R}^{2n,2n}$ such that $Q^T S Q$ is an unreduced symplectic Hessenberg matrix **if and only if** the nonlinear system of equations

$$(18.3) \qquad \begin{aligned} x^T J S^i x &= 0, \quad i = 1, \ldots, n-1 \\ x^T x &= 1 \end{aligned}$$

has a solution vector x that is not contained in an invariant subspace of S of dimension less than or equal to n.

PROOF: See Ammar/Mehrmann [A 4]. ∎

We do not know necessary and sufficient conditions for the existence of a vector satisfying (18.2) (or (18.3)), that is not contained in an invariant subspace of dimension less than or equal to n. Sufficient conditions are given in the following proposition.

18.4 PROPOSITION.

i) Let $\mathcal{H} \in \mathbf{R}^{2n,2n}$ be a Hamiltonian matrix and let $\mathcal{A}$ be an isotropic $\mathcal{H}$–invariant subspace of $\mathbf{R}^{2n}$. Then any $x \in \mathcal{A}$ such that $\| x \|_2 = 1$ satisfies conditions (18.2).

ii) Let $S \in \mathbf{R}^{2n,2n}$ be a symplectic matrix and let $\mathcal{A}$ be an isotropic S–invariant subspace of $\mathbf{R}^{2n}$. Then any $x \in \mathcal{A}$ such that $\| x \|_2 = 1$ satisfies conditions (18.3).

PROOF: Since the subspace $\mathcal{A}$ is invariant, it follows that $\mathcal{H}^k x = y \in \mathcal{A}$, $(S^k x = y \in \mathcal{A})$ for any $k = 1, \ldots, 2n$ and, since $\mathcal{A}$ is isotropic, it follows that $x^T J y = 0$. This proves i) (and ii)). ∎

Now in Section 17 we have discussed numerical methods to find the eigenvalues of Hamiltonian or symplectic matrices or pencils. Thus, if the eigenvalues are known we can apply the following Proposition:

18.5 PROPOSITION.

i) Let $\mathcal{H} \in \mathbf{R}^{2n,2n}$ be Hamiltonian and have the eigenvalues $\lambda_1, \ldots, \lambda_{2n}$ with multiplicities counted. Suppose that there exists an n-dimensional real Lagrangian invariant subspace $\mathcal{A}$ of $\mathcal{H}$ corresponding to eigenvalues $\lambda_1, \ldots, \lambda_n$ with multiplicities counted. Then

$$(\mathcal{H} - \lambda_1 I) \ldots (\mathcal{H} - \lambda_n I) e_1$$

is contained in the Lagrangian $\mathcal{H}$-invariant subspace corresponding to the eigenvalues $\lambda_{n+1}, \ldots, \lambda_{2n}$.

ii) Let $\mathcal{S} \in \mathbf{R}^{2n,2n}$ be symplectic and have the eigenvalues $\lambda_1, \ldots, \lambda_{2n}$ with multiplicities counted. Suppose that there exists an n-dimensional real Lagrangian invariant subspace $\mathcal{A}$ of $\mathcal{S}$ corresponding to the eigenvalues $\lambda_1, \ldots, \lambda_n$ with multiplicities counted. Then

$$(\mathcal{S} - \lambda_1 I) \ldots (\mathcal{S} - \lambda_n I) e_1$$

is contained in the Lagrangian $\mathcal{S}$-invariant subspace corresponding to the eigenvalues $\lambda_{n+1}, \ldots, \lambda_{2n}$.

PROOF: i) The matrix $(\mathcal{H} - \lambda_1 I) \ldots (\mathcal{H} - \lambda_n I)$ has rank less than or equal to n and is real, since the Lagrangian subspace is real, i.e., since the set $\{\lambda_1, \ldots, \lambda_n\}$ is closed under complex conjugation. Thus, there exists an orthogonal matrix Q such that

$$Q^T (\mathcal{H} - \lambda_1 I) \ldots (\mathcal{H} - \lambda_n I) = \begin{bmatrix} R_{11} & R_{12} \\ 0 & 0 \end{bmatrix}$$

is upper triangular with an $n \times n$ diagonal block R_{11}. It follows that

$$(\mathcal{H} - \lambda_1 I) \ldots (\mathcal{H} - \lambda_n I) Q = Q \begin{bmatrix} \tilde{R}_{11} & \tilde{R}_{12} \\ 0 & 0 \end{bmatrix}.$$

Now splitting Q as $[Q_1 \quad Q_2]$, $Q_1, Q_2 \in \mathbf{R}^{2n,n}$, it follows that the columns of Q_1 span the invariant subspace corresponding to the eigenvalues $-\lambda_{n+1}, \ldots, -\lambda_{2n}$. Thus, $Q_1 e_1$ is contained in this subspace, which is the Lagrangian subspace $J\mathcal{A}$. This proves i). For ii) replace $\mathcal{H}$ by $\mathcal{S}$ in the proof of i). ∎

In our applications of optimal control problems, one has a Hamiltonian matrix with exactly n eigenvalues in the left half plane and n eigenvalues in the right half plane and one is interested in computing the *stable* invariant subspace corresponding to the eigenvalues in the left half plane. But there are also examples in H_∞-control, where a Lagrangian invariant subspace of a Hamiltonian matrix with eigenvalues on the imaginary axis has to be computed, see e.g., Clements/Glover [C 8]. Such a Hamiltonian matrix has a Lagrangian invariant subspace provided that its purely imaginary eigenvalues occur with even multiplicity, see e.g. Lancaster/Rodman [L 4] or Lin [L 22].

Proposition 18.5 implies that we can compute a particular Lagrangian invariant subspace of a Hamiltonian matrix as follows:

18.6 ALGORITHM SISHC [COMPUTATION OF THE STABLE INVARIANT SUBSPACE OF A HAMILTONIAN MATRIX].

Given a Hamiltonian matrix $\mathcal{H} \in \mathbb{R}^{2n,2n}$ having an n–dimensional Lagrangian invariant subspace $\mathcal{A}$ corresponding to the eigenvalues $\lambda_1, \ldots, \lambda_n$, and a tolerance ϵ, this algorithm computes a real orthogonal symplectic matrix $Q \in \mathbb{R}^{2n,2n}$ such that the first n columns of Q span the Lagrangian subspace of A corresponding to the other eigenvalues $\lambda_{n+1}, \ldots, \lambda_{2n}$ of A.

Set $Q := I$.

Step 1: (Computation of eigenvalues.)

Compute the eigenvalues of $\mathcal{H}$. (Use for example Algorithm SQRED.) Let $\lambda_1, \ldots, \lambda_n$ be the eigenvalues corresponding to the required Lagrangian subspace.

Step 2: (Computation of the first column of the transformation matrix and reduction.)

Form

$$(18.7) \qquad x = (\mathcal{H} - \lambda_1 I) \ldots (\mathcal{H} - \lambda_n I) e_1$$

by appropriate scaling to avoid overflow, and let $Q_1 \in \mathbb{R}^{2n,2n}$ be an orthogonal symplectic matrix such that $Q_1^T x = \alpha e_1$, $\alpha = ||x||_2$. Set

$$(18.8) \qquad \mathcal{H} := Q_1^T \mathcal{H} Q_1, \; Q := QQ_1$$

and apply Algorithm 14.7 HLRED to compute an orthogonal symplectic matrix $Q_2 \in \mathbb{R}^{2n,2n}$ with $Q_2 e_1 = e_1$ such that

$$(18.9) \qquad Q_2^T \mathcal{H} Q_2 \triangleq \begin{bmatrix} \boxed{\diagdown} & \square \\ \diagdown & \boxed{\diagdown} \end{bmatrix}.$$

Set

$$(18.10) \qquad \mathcal{H} := Q_2^T \mathcal{H} Q_2 = [h_{ij}], \; Q := QQ_2.$$

Step 3: (Deflation.)

FOR $\ell = 1, 2, \ldots$

FOR $i = 1, \ldots, n$

IF $|h_{n+i,i}| < \epsilon$ THEN set $h_{n+i,i} := 0$.

END FOR

FOR $i = 2, \ldots, n$

    IF $|h_{i,i-1}| < \epsilon$ THEN set $h_{i,i-1} := 0$.

END FOR

Let $p \geq 0$ be the largest integer in $\{\ell, \ldots, n\}$ such that $h_{p+1,p} = 0$ and $h_{n+k,k} = 0$ for $k = 1, \ldots, p$. Partition $\mathcal{H}$ as

(18.11)
$$
\begin{array}{c} p\{ \\ n-p\{ \\ p\{ \\ n-p\{ \end{array}
\begin{bmatrix}
F_{11} & F_{12} & G_{11} & G_{12} \\
0 & F_{22} & G_{21} & G_{22} \\
0 & 0 & -F_{11}^T & 0 \\
0 & H_{22} & -F_{12}^T & -F_{22}^T
\end{bmatrix}
$$

IF $p = n$ THEN

    QUIT

ELSE

    Set $\mathcal{H}_{22} := \begin{bmatrix} F_{22} & G_{22} \\ H_{22} & -F_{22}^T \end{bmatrix}$,

(18.12)
$$
x_2 := (A_{22} - \lambda_1 I) \ldots (A_{22} - \lambda_n I) e_1
$$

and compute an orthogonal symplectic matrix

$$
Q_2 = \begin{bmatrix} U_2 & V_2 \\ -V_2 & U_2 \end{bmatrix} \in \mathbf{R}^{2(n-p),2(n-p)},
$$

such that $Q_2^T x_2 = a_2 e_1$. Set

(18.13)
$$
Q_e := \begin{array}{c} p\{ \\ n-p\{ \\ p\{ \\ n-p\{ \end{array}
\begin{bmatrix}
I & 0 & 0 & 0 \\
0 & U_2 & 0 & V_2 \\
0 & 0 & I & 0 \\
0 & -V_2 & 0 & U_2
\end{bmatrix}, \mathcal{H} := \tilde{Q}_e^T \mathcal{H} \tilde{Q}_e, \; Q := Q \cdot \tilde{Q}_e
$$

and apply Algorithm 14.7 HLRED to compute an orthogonal symplectic $\tilde{Q}_2 \in \mathbf{R}^{2n,2n}$ with $\tilde{Q}_2 e_1 = e_1$ such that $\tilde{Q}_2^T \mathcal{H} \tilde{Q}_2$ has then from (18.9).

    END IF

  END FOR

END

A corresponding complex version is denoted by CSISHC and easily obtained by replacing HLRED by CHLRED in the appropriate places and using CQR in Step 1 to compute the eigenvalues. Note that the other possible methods, HAMSR, SQRED do not have a complex counterpart.

If x in (18.7) has a component in each direction of the required invariant subspace A, then Algorithm 18.6 in exact arithmetic will terminate with $p = n$ immediately after Step 2. If this is not the case, then a deflation takes place and the first p columns of A span a p-dimensional invariant subspace Q_p and x_2 in (18.11) is in $A \backslash Q_p$. We can then proceed with the deflated matrix.

In the symplectic case the algorithm can be constructed in an analogous way. The analogoue of the condensed form in (18.9) is a symplectic matrix of the form

(18.14)
$$\left[\begin{array}{cc} \boxed{\diagdown} & \boxed{} \\ \boxed{\diagdown} & \boxed{} \end{array}\right],$$

which can be obtained exactly the same way. The partitioning in (18.11) becomes

(18.15)
$$\begin{bmatrix} S_{11} & S_{12} & S_{13} & S_{14} \\ 0 & S_{22} & S_{23} & S_{24} \\ S_{31} & S_{32} & S_{33} & S_{34} \\ 0 & S_{42} & S_{43} & S_{44} \end{bmatrix}$$

with S_{31} strictly upper triangular and S_{11} upper Hessenberg.
The corresponding methods for the symplectic case are denoted by SISHD for the real case and CSISHD for the complex case.
The cost of this method is $\mathcal{O}(n^3)$ flops. It depends on the number of deflation steps (which should be limited), and the method to compute the closed loop eigenvalues.

Steps 2 and 3 of this algorithm use only orthogonal symplectic transformations and thus, in exact arithmetic leave the Hamiltonian and symplectic structure invariant. Due to roundoff errors, however, these structures are only retained approximately. In the Hamiltonian case then, using the given symmetries, it is very easy to obtain a nearby Hamiltonian matrix again. In the symplectic case, however, this is not so easy. In fact we do not know how to obtain a close symplectic matrix in an easy way. Difficulties like this were already observed in other algorithms for symplectic matrices, see Sections 14, 15, 16.

First numerical tests have shown that such a method can prove to be superior to the other methods. There are still numerical difficulties in Step 2, when (18.7) is found. But since the cost of computation of this method is essentially the cost for the computation of the eigenvalues in Step 1, plus a few steps of Algorithm 14.7 HLRED, if this method is carefully coded it may be the best numerical method to compute the required invariant subspace in particular if it is combined with at least one step of defect correction, which is still advisable. Another advantage over all the other methods discussed in the previous sections is, that this method can deal also with the case of purely imaginary eigenvalues and thus, can also serve as a method to solve H_∞ control problems. Further numerical properties of this method are currently under investigation.

§ 19 Numerical algorithms for Riccati differential or difference equations.

We have so far mainly discussed algebraic Riccati equations, i.e. the case $K, T = \infty$ in the control problem. In Section 6 we have discussed how these methods can also be applied in the finite time case $K, T < \infty$ for the solution of the linear boundary value problems (3.19), (3.48). Another approach, which also works for nonconstant coefficient matrices is discussed in Kunkel/Mehrmann [K 26],[K 27]. Here we discuss only the constant coefficient case. A similar construction for this case can also be found in Bender/Laub [B 11], [B 12].

In Section 9 we have given a numerical procedure to achieve that $\alpha E - \beta A$ is regular, $\mathrm{ind}_\infty(E, A) \leq 1$, and $E = \begin{bmatrix} \Sigma_E & 0 \\ 0 & 0 \end{bmatrix}$ diagonal, with Σ_E nonsingular. If we apply the transformation in (9.8) directly to the Riccati differential equation (5.5), we obtain the system:

$$
\begin{aligned}
(19.1) \quad & -\begin{bmatrix} \Sigma_E & 0 \\ 0 & 0 \end{bmatrix}^* \begin{bmatrix} Y_{11} & Y_{12} \\ Y_{21} & Y_{22} \end{bmatrix} \begin{bmatrix} \Sigma_E & 0 \\ 0 & 0 \end{bmatrix} = \begin{bmatrix} \Sigma_E & 0 \\ 0 & 0 \end{bmatrix} \begin{bmatrix} Y_{11} & Y_{12} \\ Y_{21} & Y_{22} \end{bmatrix} \begin{bmatrix} A_{11} & A_{12} \\ A_{21} & A_{22} \end{bmatrix} \\
& + \begin{bmatrix} A_{11}^* & A_{12} \\ A_{21} & A_{22} \end{bmatrix}^* \begin{bmatrix} Y_{11} & Y_{12} \\ Y_{21} & Y_{22} \end{bmatrix} \begin{bmatrix} \Sigma_E & 0 \\ 0 & 0 \end{bmatrix} + \begin{bmatrix} Q_{11} & Q_{12} \\ Q_{21} & Q_{22} \end{bmatrix} \\
& - \begin{bmatrix} \Sigma_E & 0 \\ 0 & 0 \end{bmatrix}^* \left(\begin{bmatrix} S_{11} \\ S_{21} \end{bmatrix} + \begin{bmatrix} Y_{11} & Y_{12} \\ Y_{21} & Y_{22} \end{bmatrix} [B_{11} \; B_{21}] \right) \\
& R^{-1} \left(\begin{bmatrix} B_{11}^* \\ B_{21}^* \end{bmatrix} \begin{bmatrix} Y_{11} & Y_{12} \\ Y_{21} & Y_{22} \end{bmatrix} + \begin{bmatrix} S_{11} \\ S_{21} \end{bmatrix}^* \right) \begin{bmatrix} \Sigma_E & 0 \\ 0 & 0 \end{bmatrix}
\end{aligned}
$$

with 'terminal' condition

$$
(19.2) \qquad \Sigma_E^* Y_{11}(T) \Sigma_E = M_{11},
$$

which decouples in four equations

$$
(19.3) \qquad 0 = Q_{22}
$$

$$
(19.4) \qquad 0 = (A_{12}^* Y_{11} + A_{22}^* Y_{21}) \Sigma_E + Q_{21}
$$

$$
(19.5) \qquad 0 = \Sigma_E (Y_{12} A_{22} + Y_{11} A_{12}) + Q_{12}
$$

$$
\begin{aligned}
(19.6) \quad & - \Sigma_E \dot{Y}_{11} \Sigma_E = \Sigma_E (Y_{11} A_{11} + Y_{12} A_{21}) + (A_{11}^* Y_{11} + A_{21}^* Y_{21}) \Sigma_E \\
& + Q_{11} - \Sigma_E (S_{11}^* + B_{11}^* Y_{11} + B_{21}^* Y_{21})^* R^{-1} (S_{11}^* + B_{11}^* Y_{11} + B_{21}^* Y_{21}) \Sigma_E.
\end{aligned}
$$

Solving (19.4), (19.5) for Y_{12}, Y_{21} and substituting in (19.6) yields again a Riccati differential equation. Applying the theory of Section 2 we obtain

$$
(19.7) \qquad u(t) = -R^{-1} \left([B_{11}^* B_{21}^*] \begin{bmatrix} Y_{11} & Y_{12} \\ Y_{21} & Y_{22} \end{bmatrix} \begin{bmatrix} \Sigma_E & 0 \\ 0 & 0 \end{bmatrix} + \begin{bmatrix} S_{11} \\ S_{21} \end{bmatrix}^* \right) \begin{bmatrix} x_1(t) \\ x_2(t) \end{bmatrix}
$$

as the solution of the optimal control problem.

Thus, we could also work directly with the Riccati equation rather than performing the reduction in advance. More important, however, in this analysis is that the components in Y_{22} are not occuring in the optimal control, so they may be arbitrary. We may therefore employ a modified solution method for differential algebraic equation to solve (19.1). Such methods are well studied and implemented in packages like LIMEX of Deuflhard et al [D 9], see also Brenan et al [B 12].

In Kunkel/Mehrmann [K 26] the following Algorithm is given:

19.8 ALGORITHM DAREC.

Given $A, E, M, Q, W, M \in \mathbf{C}^{n,n}$ constant with $\alpha E - \beta A$ regular and $\mathrm{ind}_\infty(E, A) \leq 1$, $Q = Q^T$, $W = W^T$, $M = M^T$, t_0, $T \in \mathbf{R}$ and a rank tolerance $\epsilon \in \mathbf{R}$, this algorithm solves the Riccati differential equation (5.5) in the interval $[t_0, T]$.

Step 1

Perform a singular value decomposition of E using the LINPACK DSVDC routine, i.e. determine $U, V \in \mathbf{C}^{n,n}$ unitary, such that $U^* E V = \Sigma_0 = \mathrm{diag}(\sigma_1, \ldots, \sigma_n)$, $\sigma_1 \geq \sigma_2 \geq \ldots \geq \sigma_n \geq 0$.

Step 2

Determine the rank p of Σ_0 as the largest index $i \in \{1, \ldots, n\}$, such that $\sigma_i \geq \epsilon \sigma_1$ and set $\sigma_{p+1}, \ldots, \sigma_n$ to zero.

Step 3

Perform the transformation with U, V to the whole Riccati equation, i.e. set $A := U^* A V$, $Q := V^* Q V$, $W := U^* W U$, $Y(t) := U^* X(t) U$, $M := V^* M V$, $E := \begin{bmatrix} \Sigma_E & 0 \\ 0 & 0 \end{bmatrix}$, where $\Sigma_E = \mathrm{diag}\{\sigma_1, \ldots, \sigma_p\}$. Partition the matrices analogous to the partitioning in E.

Step 4 Check the consistency conditions

$$Q_{22} = 0, \ M_{12} = 0, \ M_{22} = 0.$$

If these conditions do not hold, the system is not solvable,

STOP

Step 5

Solve the algebraic equation (19.5) for $Y_{12}(T)$, i.e. for the "terminal" value T using the QR-decomposition with column pivoting for A_{22}, e.g. Golub/Van Loan [G 12].

(19.8) $Y_{12}(T) = -(Y_{11}(T)A_{12} + \Sigma_E^{-1}Q_{12})A_{22}^{-1} = -\Sigma_E^{-1}(M_{11}\Sigma_E^{-1}A_{12} + Q_{12})A_{22}^{-1}.$

Step 6

Rewrite the relevant part of $Y(t) = \begin{bmatrix} Y_{11} & Y_{12} \\ Y_{21} & Y_{22} \end{bmatrix}$ which is given by the shaded region

in: [figure], i.e. the upper triangle of Y_{11} including the diagonal and Y_{12}, in

vector form, by going through $[Y_{11} \quad Y_{12}]$ rowwise. Let

$[z_{11}, \ldots, z_{1n}, z_{22}, \ldots, z_{2n}, \ldots, z_p, \ldots, z_{pn}]^T =: v =: [v_1, \ldots, v_\ell]^T$, where $\ell = \frac{p(2n-p+1)}{2}$,

and rewrite system (19.6) in terms of v.

Call the new system

(19.9) $$\tilde{E}\dot{v}(t) = \tilde{f}(t, v(t)).$$

Step 7

Solve system (19.9) by calling LIMEX of Deuflhard et al [D 9] (or any other system for differential algebraic equations).

Step 8

Rewrite the solution v as a matrix by completing the matrix $Y(t_0) = \begin{bmatrix} Y_{11}(t_0) & Y_{12}(t_0) \\ Y_{21}(t_0) & Y_{22}(t_0) \end{bmatrix}$ with 0 components in Y_{21}, and in a symmetric way in Y_{21}.

Step 9 Form $X(t_0) = U \, Y(t_0) U^T$.

For more details, numerical examples and also the time varying case see Kunkel/Mehrmann [K 26], [K 27].

Observe that if we have already preprocessed the system as in Section 9, then Steps 1, 2, 3 can be omitted, and A_{22} is already diagonal, so Step 5 is also much easier and essentially only steps 6,7,8 have to be performed.

In the discrete case the reduction can be performed as in Section 9, which leads to the control problem

(19.10) $$\begin{bmatrix} \Sigma_E & 0 \\ 0 & 0 \end{bmatrix} \begin{bmatrix} x_{k+1}^{(1)} \\ x_{k+1}^{(2)} \end{bmatrix} = \begin{bmatrix} A_{11} & A_{12} \\ A_{21} & A_{22} \end{bmatrix} \begin{bmatrix} x_k^{(1)} \\ x_k^{(2)} \end{bmatrix} + \begin{bmatrix} B_{11} \\ B_{21} \end{bmatrix} u_k$$

$$\begin{bmatrix} x_0^{(1)} \\ x_0^{(1)} \end{bmatrix} = x^0, \; y_k = [C_{11} \; C_{12}] \begin{bmatrix} x_k^{(1)} \\ x_k^{(2)} \end{bmatrix}$$

with cost functional

(19.11) $$S(x_k, u_k) = \frac{1}{2}(x_K^* M x_K + \sum_{k=k_0}^{K-1}(x_k^* Q x_k + x_k^* S u_k + u_k^* S^* x_k + u_k^* R u_k))$$

and M, Q, S, R are as in (9.9). Thus, by assumption $M = \begin{bmatrix} M_{11} & 0 \\ 0 & 0 \end{bmatrix}$.

In (19.10) we can solve for $x_k^{(2)}$ and get

(19.12) $$x_k^{(2)} = -A_{22}^{-1}A_{21}x_k^{(1)} - A_{22}^{-1}B_{21}u_k$$

which gives the consistency condition for u_{k_0}. Substituting (19.12) in (19.10), (19.11) yields the system

(19.13) $\Sigma_E x_{k+1}^{(1)} = (A_{11} - A_{12}A_{22}^{-1}A_{21})x_k^{(1)} + (B_{11} - A_{12}A_{22}^{-1}B_{21})u_k =: \tilde{A}_{11}x_k^{(1)} + \tilde{B}_{11}u_k$

and the cost functional
(19.14)
$$S(x^{(1)}, u) = \frac{1}{2}\Big[x_K^{(1)*}M_{11}x_K^{(1)} +$$

$$\sum_{k=k_0}^{K-1}\Big((x_k^{(1)*}(Q_{11} - S_{12}^*A_{22}^{-1}A_{21} - A_{21}^*A_{22}^{-1}S_{21} - Q_{12}A_{22}^{-1}A_{21} - A_{21}^*A_{22}^{-*}Q_{21} +$$

$$A_{21}^*A_{22}^{-*}Q_{12}A_{22}^{-1}A_{21})x_k^{(1)} + x_k^{(1)*}(S_{11} - A_{21}^*A_{22}^{-1}S_{21})u_k + u_k^*(S_{11}^* - S_{21}^*A_{22}^{-*}A_{21})x_k^{(1)} +$$

$$u_k^*(R + B_{21}^*A_{22}^{-*}Q_{21}A_{22}^{-1}B_{21} - S_{21}^*A_{22}^{-1}B_{21} - B_{21}^*A_{22}^{-*}S_{21})u_k)\Big]$$

$$=: \frac{1}{2}\Big[x_K^{(1)}M_{11}x_K^{(1)} + \sum_{k=k_0}^{K-1}(x_k^{(1)*}\tilde{Q}_{11}x_k^{(1)} + x_k^{(1)*}\tilde{S}_{11}u_k + u_k^*\tilde{S}_{11}^*x_k^{(1)} + u_k^*\tilde{R}u_k)\Big].$$

The optimal solution (if it exists) is thus

(19.15) $$u_k = -(\tilde{R} + \tilde{B}_{11}^*X_kB_{11})^{-1}(\tilde{A}_{11}^*X_kB_{11} + \tilde{S}_{11})^*x_k^{(1)}$$

where X_k solves the Riccati difference equation
(19.16)
$$\Sigma_E^*X_kE = \tilde{Q}_{11} + \tilde{A}_{11}^*X_{k+1}\tilde{A}_{11}$$
$$- (\tilde{A}_{11}^*X_{k+1}\tilde{B}_{11} + \tilde{S}_{11})(\tilde{R} + \tilde{B}_{11}^*X_{k+1}\tilde{B}_{11})^{-1}(\tilde{A}_{11}^*X_k\tilde{B}_{11} - \tilde{S}_{11})^*, \ k = k_0, \ldots k-1$$

with terminal condition

(19.17) $$\Sigma_E X_K \Sigma_E = M_{11}.$$

This system can now be solved step by step by solving linear systems with $\tilde{R} + \tilde{B}_{11}^*X_{k+1}\tilde{B}_{11}$, which can be done via the Cholesky factorization, e.g. Golub/Van Loan [G 12], at each step. Depending on the number of steps that have to be performed it may be cheaper to use the deflating subspace method of Section 6, for the boundary value problem (3.48) directly. We denote this procedure by DARECD. See also Nikoukhah et al [N 1] for a discussion of this topic in the time varying case.

Also if A_{22}, R are ill–conditioned then working directly with the boundary problems (3.11), (3.48) is more advisable in order to get accurate results.

In this Section we have discussed the numerical solution of the Riccati differential or difference equation as a method to solve the control problems with a different approach if $K, T < \infty$. In the case of nonconstant coefficient matrices this is essentially the only way to go, e.g. Kunkel/Mehrmann [K 26], [K 27], but for constant coefficients the subspace approach may prove much more efficient.

§ 20 A general algorithm.

We now combine the algorithms described in Sections 9–19 to one 'expert system' for the solution of linear quadratic optimal control problems (1.1), (1.2). The general algorithm for each of the problems has the form of a decision flow chart. It is possible to join the two diagrams into one, in particular since the top part of the two charts is identical, but this would have made it very difficult to display, so we have separated the two problems. The input data of the general algorithm are the system matrices A, B, C, E, M, Q, R, S, their dimensions p, m, n as described in (1.1), the initial state x^0 and the initial and final times t_0, k_0, T, K. The diagrams are partitioned into 8 levels. The first 4 levels serve the purpose of testing system properties and simplifying the system, mainly by carrying out the preprocessing procedures discussed in Section 9. In particular we distinguish here between the real and complex case and also whether $E = I$ or not. As we have seen in Section 9, it is possible to test properties of the system that we have assumed throughout this book, like constraints on M as discussed in Remark 3.16, condition numbers of matrices that have to be inverted, like cond (Σ_E), cond (A_{22}) or cond (R), or conditions which violate strong stabilizability and detectability. (See Section 9.)

If $E \neq I$ then we also have to ensure that the pencil $\alpha E - \beta A$ is regular and has index ≤ 1. If this is not possible or if some of the constraints are not satisfied, (all this is tested within or following CREGSCD,) then we have to stop, since the considered problem is not solvable. This is tested in level 1 with the decision $< E, A, B, C, Q, M$ ok $>$ and this could be implemented directly into the system. In levels 2 to 4 it is possible to implement the decisions whether matrices are wellconditioned (WC) either with interrupts, which allow the user to steer the system, or by automatic decisions.

In level 5 it is distinguished between the infinite horizon ($T, K = \infty$) and the finite horizon problem ($T, K < \infty$). In the finite horizon problem we have the choice between the subspace method, see Section 6, or DARE solvers, see Section 19. Either choice yields a reliable method. It is difficult to decide which of the methods is more efficient, since the efficiency depends on the sizes of m, n, p, the rank of E and the number of integration steps needed in a DARE solver. If the rank of E is small compared to n, then the DARE solver DAREC is relatively cheap compared with the computation of the generalized Schur form of $\alpha A - \beta B$, or $\alpha A' - \beta B'$.

Also on this level we test for $rk(G) = 1$ or $rk(H) = 1$. This decision is not based on an SVD but we make here use of a priori knowledge about the system. These rank conditions are automatically given in single output or single input systems and we base this decision on this fact.

20.1 CHART. *Continuous time problem (1.1)*

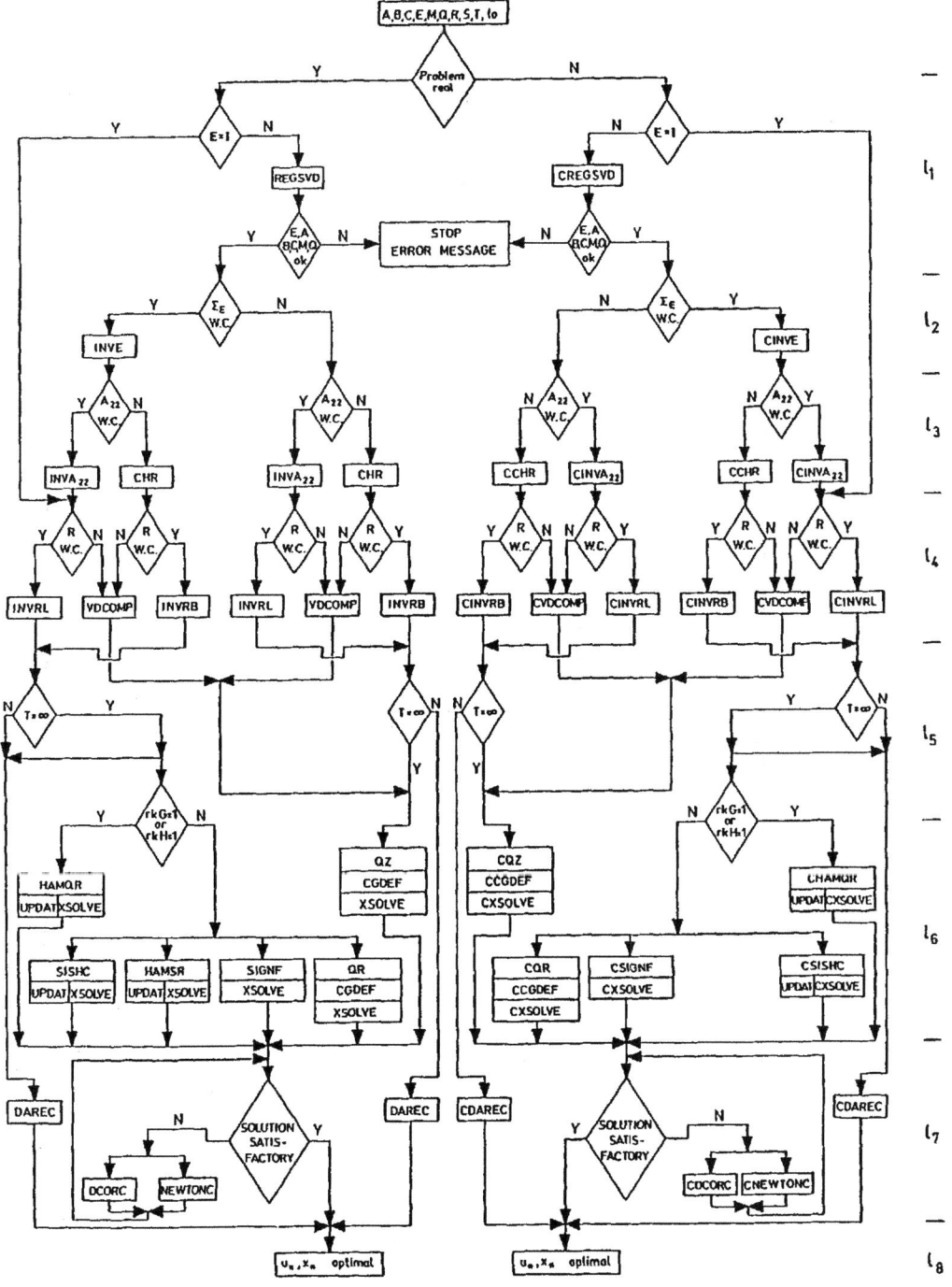

154

20.2 CHART. *Discrete time problem (1.2)*

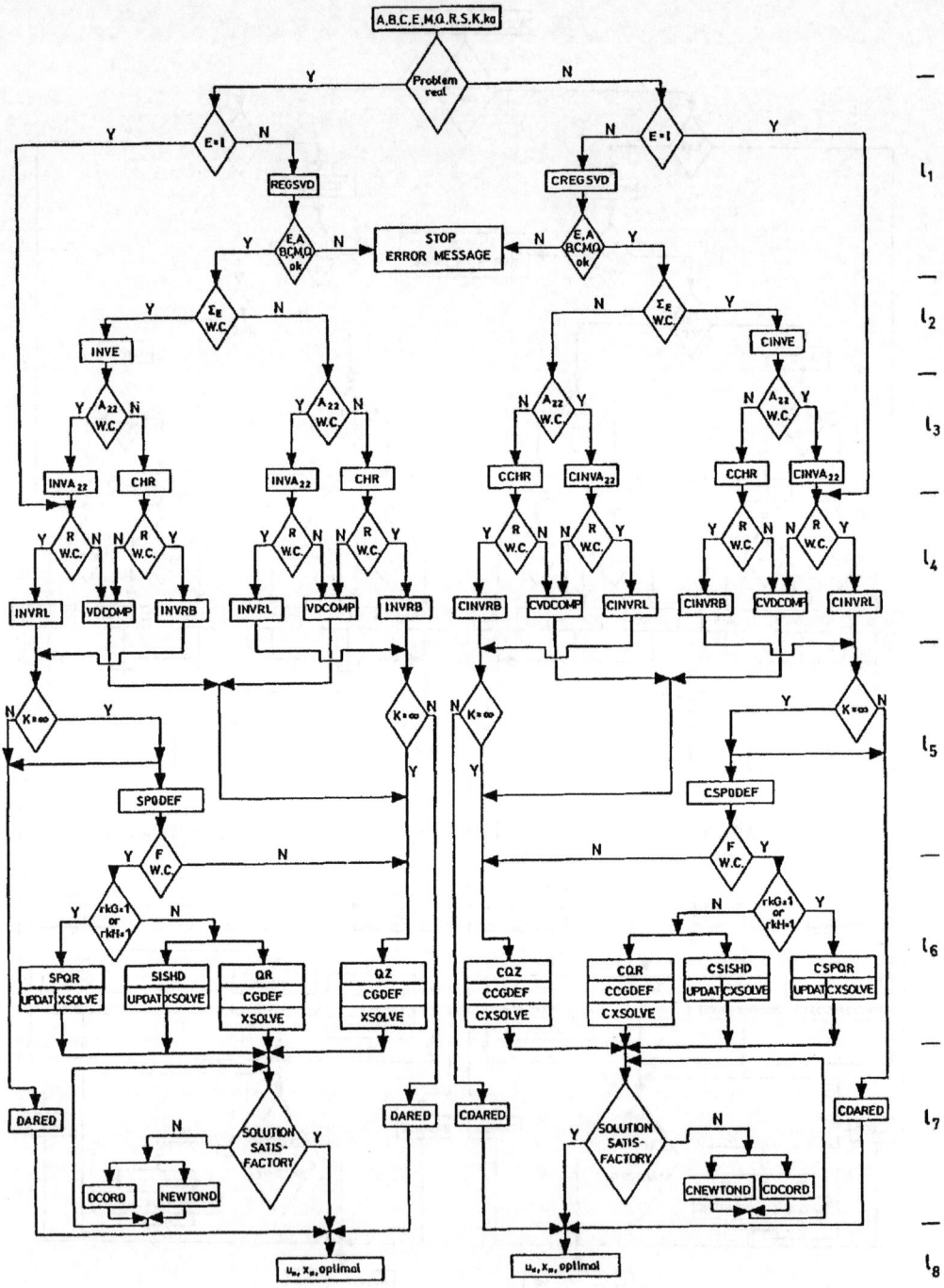

If the rank test yields **Y**, then the best choice are the methods HAMQR, CHAMQR, SPQRP, SPQR, CSPQRP, CSPQR discussed in Section 15, since they are numerically (strongly) stable, preserve the structure and are usually more efficient than any of the other solvers discussed on this level.

In the previous sections we have discussed several methods to compute different Schur forms of Hamiltonian or symplectic matrices or symplectic pencils, from which we then can determine the deflating subspace that serves to compute the Riccati solution or the optimal control directly. Having determined this subspace, we need the first block component to be transformed to be the identity matrix, i.e. if $\begin{bmatrix} S_1 \\ S_2 \\ S_3 \end{bmatrix} \in K_{2n+m,n}$ are the first n columns of the transformation matrix, which span the deflating subspace in the case where R is not inverted, or $\begin{bmatrix} S_1 \\ S_2 \end{bmatrix} \in K_{2n,n}$ are the first n columns of the transformation matrix otherwise, then we have seen that the required solution of the Riccati equation is $XE = -S_2 S_1^{-1}$. This solution can be obtained via Gaussian elimination with pivoting or via the QR-decomposition. A procedure which performs this inversion is called CXSOLVE, (XSOLVE). In some of the methods it can be replaced by a direct updating of the solution denoted by UPDAT, see Section 16. This alternative we denote by splitting the box corresponding to one method in the lower part, which means that both alternatives are possible.

If neither of the rank tests in level 6 yields **Y**, then we have several alternatives to chose from, none of which is completely satisfactory.

Either the method is not preserving the (Hamiltonian or symplectic) structure: QR, CQR, QZ, CQZ combined with the deflation procedure of Clements/Glover CCGDEF (CGDEF), see Section 14, or the method is not numerically backwards stable, CSIGNF, (SIGNF), HAMSR, see Sections 11, 16.

Probably the best choice is to use the combination algorithms CSISHD, CSISHC, SISHD, SISHC discussed in Section 18, but for a definite statement on this topic more numerical experience is needed, which is currently not available.
Operation counts for the different methods on this level show no significant difference between the methods, they are all $\mathcal{O}(n^3)$ methods, with different factors though. See Sections 14, 15, 16, 17, 18.

In level 7 we have the possibilty to decide whether we are satisfied with the accuracy of the solution and we can use iterative refinement to improve the accuracy if necessary. Here, there are many alternatives to choose a method for the refinement step. Numerical experience seems to indicate that Newtons method or any other method in the defect correction algorithm are equally good choices, provided the initial solution has at least some accurate digits, see Sections 10,11.

In level 8, then the optimal solution and optimal trajectory is computed. Having computed X or often even XE, we can use formulae (1.18), (1.21) to compute the optimal feedback. These formulae can often be further simplified by using previously computed products.

§21 Conclusion.

In this book we have given a survey of the state of the art on the theory and the numerical solution of control problems (1.1), (1.2). Many of the topics discussed here are still areas of very active research, so some of the discussed subjects are given in more detail then others. We apologize for every recent or old reference that we have overlooked.

We hope that the attempt to join together the theoretical mathematical results as well as the numerical algorithms helps researchers of both sides as well as people who work in applications to get deeper into this fascinating subject.

We would have very much liked to accompany this book with a software package that contains all the discussed procedures. This is a plan for the future, but so far we can only advice the reader to look for the bits and pieces in the given references.

We also hope that we initiate with this book a closer cooperation between engineers, pure mathematicians and numerical analysts. For this reason we have tried to use mainly elementary mathematical tools and also have tried to avoid details of applications, while keeping the problem as general and widely applicable as possible. Further generalizations are referenced in the corresponding sections.

§22 Statement.

We are completely aware of the fact that some the of major applications of optimal control are in the construction and use of military equipment. We cannot avoid the use of the described theoretical and numerical results in these military applications, but we do not approve of these applications.

§23 References.

[A 1] J.C. Allwright, *A lower bound for the solution of the algebraic Riccati equation of optimal control and a geometric convergence rate for the Kleinman algorithm*, IEEE Trans. Autom. Control, AC–25 (1980), 826–829.

[A 2] G.S. Ammar, *Riccati equations and the numerical matrix eigenvalue method*, PhD thesis, Case Western Reverse Univ. (1983).

[A 3] G.S. Ammar, C. Martin, *The geometry of matrix eigenvalue methods*, Acta Appl. Math. 5 (1986), 239–278.

[A 4] G.S. Ammar, V. Mehrmann, *On Hamiltonian and symplectic Hessenberg forms*, Lin. Alg. Appl. 149 (1991), 55–72.

[A 5] B.D.O. Anderson, *Second order convergent algorithms for the steady–state Riccati equation*, Int. J. Control 28 (1978), 295–306.

[A 6] B.D.O. Anderson, J.B. Moore, *Linear Optimal Control*, Prentice Hall, Englewood Cliffs, N.J. (1971).

[A 7] B.D.O. Anderson, J.B. Moore, *Optimal Filtering*, Prentice Hall, Englewood Cliffs, New Jersey (1979).

[A 8] B.D.O. Anderson, B. Vongpanitlerd, *Network analysis and synthesis. A Modern Systems Approach*, Prentice Hall, Englewood Cliffs, N.J. (1972).

[A 9] F.W. Arnold III, *On the numerical solution of algebraic Riccati equations*, Ph.D. Thesis, Dept. of Electrical Engineering, Univ. of Southern California (1983).

[A 10] F.W. Arnold III, *Numerical solution of algebraic Riccati equations*, Technical Rep., Naval Weapons Research Center, China Lake NWCTP 6521 (1984).

[A 11] F.W. Arnold III, A.J. Laub, *A software package for the solution of generalized algebraic Riccati equations*, Proc. 22nd CDC, San Antonio (Dec. 1983).

[A 12] F.W. Arnold III, A.J. Laub, *Generalized eigenproblem algorithms and software for algebraic Riccati equations*, Proc. of IEEE 72 (1984), 1746–1754.

[A 13] E. Artin, *Geometric Algebra*, Interscience, New York (1957).

[A 14] M. Athans, *The role and use of the stochastic linear quadratic Gaussian problem in control system design*, IEEE Trans. Autom. Control, AC–16 (1971), 529–552.

[A 15] M. Athans, P.L. Falb, *Optimal Control*, Mac Graw Hill, New York (1966).

[B 1] L. Balzer, *Accelerated convergence of the matrix sign function*, Int. J. Control, 21 (1980), 1057–1078.

[B 2] Y. Bar–Ness, *Sufficient conditions for the solution of the discrete time linear regulator*, Int. J. Control 24 (1976), 335–343.

[B 3] A.Y. Barraud, *Investigation autour de la fonction signe d'une matrice, application á l'équation de Riccati*, R.A.I.R.O. Automatique 13 (1979), 335–368.

[B 4] A.Y. Barraud, *Produit étoile et fonction signe de matrice. Application a l'équation de Riccati dans le cas discret*, R.A.I.R.O. Automatique 14 (1980), 55–85.

[B 5] R.H. Bartels, G.W. Stewart, *Solution of the matrix equation $AX + XB = C$*, Algorithm 432, Comm. ACM-15 (1972), 820–826.

[B 6] A.N. Beavers, E.D. Denman, *A computational method for eigenvalues and eigenvectors of a matrix with real eigenvalues*, Numer. Math. 21 (1974), 389–396.

[B 7] A.N. Beavers, E.D. Denman, *Asymptotic solutions to the matrix Riccati equation*, Mathematical Biosciences 20 (1974), 339–344.

[B 8] A.N. Beavers, E.D. Denman, *A new solution method for matrix quadratic equations*, Mathematical Biosciences 20 (1974), 135-143.

[B 9] A.N. Beavers, E.D. Denman, *A new similarity transformation method for eigenvalues and eigenvectors*, Mathematical Biosciences 21 (1974), 143–169.

[B 10] J. G. F. Belinfante, B. Kolman, *A Survey of Lie Groups and Lie Algebras with Applications and Computational Methods*, SIAM, Philadelphia, PA (1972).

[B 11] D.J. Bender, A.J. Laub, *The linear quadratic optimal regulator for descriptor system: Discrete-Time case*, Automatica, 23 (1987), 71–86.

[B 12] D.J. Bender, A.J. Laub, *The linear-quadratic optimal regulator for descriptor systems*, IEEE Trans. Autom. Control AC-32 (1987), 672–688.

[B 15] J.P. Birdwell, A.J. Laub, *Balanced singular values for LQG/LTR design*, Int. J. Control 45 (1987), 939–950.

[B 16] S. Bittanti, P. Bolzern, *On the structure theory of discrete time linear systems*, Int. J. Systems Sci. 17 (1986), 33–47.

[B 17] K.E. Brenan, S.L. Campbell, L.R. Petzold, *Numerical Solution of Initial-Value Problems in Differential-Algebraic Equations*, Elsevier Science Publishing, North-Holland (1989).

[B 18] R.A. Brocket, *Finite Dimensional Linear Systems*, Wiley, New York (1970).

[B 19] R.S. Bucy, *Global theory of the Riccati equation*, J. Comput. Syst. Sci. 1 (1967), 349–361.

[B 20] R.S. Bucy, *The Riccati equation and its bounds*, J. Comput. Syst. Sci. 6 (1972), 343–353.

[B 21] R.S. Bucy, *Structural stability for the Riccati equation*, SIAM J. Control 13 (1975), 749–753.

[B 22] J.R. Bunch, *The weak and strong stability of algorithms in numerical algebra*, Lin. Alg. Appl. 88 (1987), 49–66.

[B 23] W. Bunse, A. Bunse–Gerstner, *Numerische Lineare Algebra*, Teubner, Stuttgart (1985).

[B 24] A. Bunse–Gerstner, *An algorithm for the symmetric generalized eigenvalue problem*, Lin. Alg. Appl. 58 (1984), 43–68.

[B 25] A. Bunse–Gerstner, *QR like algorithms*, Habilitationsschrift, Universität Bielefeld (Juli 1986).

[B 26] A. Bunse–Gerstner, *Eigenvalue algorithms for matrices with special structure*, Colloquia Math. Soc. János Bolyai 50. Numerical Methods, Miskolc (Hungary) (1986), 141–163.

[B 27] A. Bunse–Gerstner, *Matrix factorization for symplectic QR–like methods*, Lin. Alg. Appl. 83 (1986), 49–77.

[B 28] A. Bunse–Gerstner, V. Mehrmann, *The HHDR-algorithm and its application to optimal control problems*, R.A.I.R.O. Automatique 23 (1989), 305–329.

[B 29] A. Bunse–Gerstner, V. Mehrmann, *A symplectic QR–like algorithm for the solution of the real algebraic Riccati equation*, IEEE Trans. Autom. Control AC–31 (1986), 1104–1113.

[B 30] A. Bunse–Gerstner, R. Byers, V. Mehrmann, *A chart of numerical methods for structured eigenvalue problems*, to appear in SIAM Journal on Matrix Analysis and Applications.

[B 31] A. Bunse–Gerstner, R. Byers, V. Mehrmann, *The quaternion QR algorithm*, Numer. Math. 55 (1989), 83–95.

[B 32] A. Bunse–Gerstner, R. Byers, V. Mehrmann, *Numerical methods for algebraic Riccati equations*, Lecture Notes of the Workshop on "The Riccati Equation in Control, Systems and Signal", S. Bittanti, Ed., Pitagora Editrice Bologna (1989), 107–115.

[B 33] A. Bunse–Gerstner, V. Mehrmann, N.K. Nichols, *Derivative feedback for descriptor systems*, FSP Mathematisierung, Universität Bielefeld, Materialien LVIII (1989).

[B 34] A. Bunse–Gerstner, V. Mehrmann, N.K. Nichols, *On derivative and proportional feedback design for descriptor systems*, Proceedings of the International Symposium on the Mathematical Theory of Networks and Systems, M.A. Kaashoek et al, Edts., Amsterdam (1989).

[B 35] A. Bunse–Gerstner, V. Mehrmann, N.K. Nichols, *Regularization of descriptor systems by derivative and proportional state feedback*, Report 3/91 Numerical Analysis Group, Dept. of Math. University of Reading (1991).

[B 36] A. Bunse–Gerstner, V. Mehrmann, D. Watkins, *An SR algorithm for Hamiltonian matrices, based on Gaussian elimination*, Methods of Operations Research 58 (1989), 15–26.

[B 37] N. Burgoyne, R. Cushman, *Normal forms for real linear Hamiltonian systems*, in: C. Martin, R. Herman (eds.): The 1976 Ames Research Center (NASA) Conference on Geometric Control Theory, Mathematical Science Press, Brookline, Mass. (1976), 483–528.

[B 38] R.S. Burns, *Application of the Riccati equation in the control and guidance of marine vehicles*, Lecture Notes of the Workshop on "The Riccati Equation in Control, Systems and Signal", S. Bittanti, Ed., Pitagora Editrice Bologna (June 1989), 18–23.

[B 39] R. Byers, *Hamiltonian and symplectic algorithms for the algebraic Riccati equation*, Ph.D. Thesis Cornell Univ. (Jan. 1983).

[B 40] R. Byers, *Numerical condition of the algebraic Riccati equation*, Lin. Alg. and its Role in Systems theory, Cont. Math. 47, AMS (1984), 35–51.

[B 41] R. Byers, *A Hamiltonian QR-algorithm*, SIAM J. Sci. Stat. Comp. 7 (1986), 212–229.

[B 42] R. Byers, *A Hamiltonian-Jacobi algorithm*, to appear in IEEE Trans. Autom. Control.

[B 43] R. Byers, *Solving the algebraic Riccati equation with the matrix sign function*, Lin. Alg. Appl. 85 (1987), 267–279.

[B 44] R. Byers, *Numerical stability and instability in matrix sign function based algorithms*, Comp. and Comb. Methods in Systems Theory, C.I. Byrnes, A. Lindquist, Edts, North Holland, New York (1986), 185–200.

[B 45] R. Byers, V. Mehrmann, *Symmetric updating of the solution of the algebraic Riccati equation*, Proc. of X. Symp. on Oper. Res. München, Methods of Operations Research 54 (1986), 117–125.

[C 1] S.L. Campbell, *Optimal control of autonomous linear processes with singular matrices in the quadratic cost functional*, SIAM J. Control and Optim. 14 (1976), 1092–1106.

[C 2] S.L. Campbell, *Singular Systems of Differential Equations*, Pitman, San Francisco (1980).

[C 3] S.L. Campbell, *Consistant initial conditions for linear time varying singular systems in frequency domain and state space methods for linear systems*, C.L. Byrnes, A. Lindquist Edts., Elsevier, N. Holland (1986), 313–318.

[C 4] J.L. Casti, *Dynamical systems and their applications: Linear Theory*, Academic Press, New York (1977).

[C 5] J.L. Casti, *The linear-quadratic control problem. Some recent results and outstanding problems*, SIAM Review 22 (Oct. 1980), 459–485.

[C 6] M.J. Chapman, D.W. Pearson, D.N. Shields, *Partial singular value assignment in the design of robust observers for discrete time descriptor systems*, IMA J. Math. Control & Inf. 5 (1988), 203–213.

[C 7] A. Ciampi, *Classification of Hamiltonian linear systems*, Indiana Univ. Math. J. 23 (1973), 513 526.

[C 8] D. Clements, K. Glover, *Spectral transformations via Hermitian pencils*, Lin. Alg. Appl. 123 (1989), 797–846.

[C 9] D. Cobb, *Descriptor variable systems and optimal state regulation*, IEEE Trans. Autom. Control AC–28 (1983), 601–611.

[C 10] D.J. Cobb, *Controllability, observability and duality in singular systems*, IEEE Trans. Autom. Control AC–29 (1984), 1076–1082.

[C 11] W.A. Coppel, *Matrix quadratic equations*, Bull. Austral. Math. Soc. 10 (1974), 377–401.

[D 1] J. Della Dora, *Numerical linear algorithms and group theory*, Lin. Alg. Appl. 10 (1975), 267–283.

[D 2] J. Della Dora, *Sur quelque algorithmes de recherche de valeurs propre*, Thése soutenue á la Faculté des Sciences de Grenoble (1973).

[D 3] J.W. Demmel, *Three methods for refining estimates of invariant subspaces*, Computing 38 (1987), 43–57.

[D 4] J.W. Demmel, *On condition numbers and the distance to the nearest ill-posed problem*, Numer. Math. 51 (1987), 251–289.

[D 5] J.W. Demmel, B. Kågström, *Computing stable eigendecompositions of matrix pencils*, Lin. Alg. Appl. 88 (1987), 139–186.

[D 6] J.W. Demmel, B. Kågström, *Stably computing the Kronecker structure and reducing subspaces of singular pencils $A - \lambda B$ for uncertain data*, in Large Scale Eigenvalue problems, J. Cullum and R.A. Willoughby Edts., Elsevier, North Holland (1986), 283–323.

[D 7] E. Denman, R. Beavers, *The matrix sign function and computations in systems*, Appl. Math. Comp. 2 (1976), 63–94.

[D 8] C. De Souza, M. Gevers, G. Goodwin, *Riccati equations in optimal filtering of nonstabilizable systems having singular state transition matrices*, IEEE Trans. Autom. Control AC–31 (1986), 931–838.

[D 9] P. Deuflhard, E. Hairer, J. Zugck, *One-step and extrapolation methods for differential algebraic equations*, Numer. Math. 51 (1987), 501–516.

[D 10] L. Dieci, Y.M. Lee, R.D. Russel, *Iterative methods for solving algebraic Riccati equations*, Technical report, Dept. Math. & Stat., Simon Fraser University, Burnaby, Cananda (1988).

[D 11] J. Dongarra, J.R. Bunch, C. Moler, G.W. Stewart, *Linpack Users' Guide*, SIAM Philadelphia (1979).

[D 12] J. Dongarra, C. Moler, J.H. Wilkinson, *Improving the accuracy of computed eigenvalues and eigenvectors*, SIAM J. Num. Anal. 20 (1983), 46–58.

[D 13] P. Dorato, A. Levis, *Optimal linear regulators: The discrete time case*, IEEE Trans. Autom. Control AC–16 (1971), 613–620.

[D 14] R.C. Dorf, *Modern Control Systems*, Second Ed., Addison Wesley, Reading (1974).

[E 1] L. Elsner, *Neuere Verfahren zur Bestimmung der Eigenwerte von Matrizen*, Ansorge, Glashoff, Werner Edts., Numerische Mathematik, Birkhäuser, Basel (1979).

[E 2] L. Elsner, *On some algebraic problems in connection with general eigenvalue algorithms*, Lin. Alg. Appl. 26 (1979), 123–138.

[E 3] L. Elsner, *Private communication*, (proof for Newton's method).

[E 4] L. Elsner, *Matrix decompositions, symmetries and eigenvalue algorithms*, Proc. Matrix Theory Conf. Auburn, Alabama (1989).

[E 5] A. Emami-Naeini, *Application of the generalized eigenstructure problem to multi-variable systems and the robust servomechanism for a plant which contains an implicit internal model*, Ph.D. Thesis, Stanford University (1981).

[E 6] A. Emami-Naeini, G. Franklin, *Comments on the numerical solution of the discrete time algebraic Riccati equation*, IEEE Trans. Autom. Control AC–25 (1980), 1015–1016.

[E 7] A. Emami–Naeini, G. Franklin, *Deadbeat control & tracing of discrete–time systems*, IEEE Trans. Autom. Control AC–27 (1982), 176–181.

[E 8] G. von Escherich, *Die zweite Variation der einfachen Integrale*, Wiener Sitzungs-berichte 8 (1898), 1191–1250.

[F 1] A.F. Fath, *Computational aspects of the linear optimal regulator problem*, IEEE Trans. Autom. Control AC–14 (1969), 547–550.

[F 2] U. Flaschka, V. Mehrmann, D. Zywietz, *An analysis of structure preserving methods for symplectic eigenvalue problems*, RAIRO APII 25 (1991), 165–190.

[F 3] L.R. Fletcher, J. Kautsky, N.K. Nichols *Eigenstructure assignment in descriptor systems*, IEEE Trans. Autom. Control AC–31 (1986), 1138–1141.

[F 4] J. Francis, *The QR-transformation, Part I*, Comput. J. 4 (1961), 265–271.

[F 5] J. Francis, *The QR-transformation, Part II*, Comput. J. 5 (1962), 332–345.

[F 6] C. Führer, *Differentiell–Algebraische Gleichungssysteme in mechanischen Mehrkör-persystemen, Theorie, Numerische Ansätze und Anwendungen*, Preprint Technische Universität München, TUM-M8807 (1988).

[G 1] G.J. Gaalman, *Comments on a nonrecursive algebraic solution for the discrete Riccati equation*, IEEE Trans. Autom. Control AC–25 (1980), 610–612.

[G 2] B.S. Garbow, J.M. Boyle, J.J. Dongarra, C.B. Moler, *Matrix Eigensystem Rou-tines-EISPACK Guide Extension*, Springer, New York (1977).

[G 3] J.P. Gahinet, A.J. Laub, *Computable bounds for the sensitivity of the algebraic Ric-cati equation*, Scientific Computation Laboratory, Technical report SCL 89–10, University of California of Santa Barbara (May 1989).

[G 4] F.R. Gantmacher, *Theory of Matrices*, Vol I Chelsea, New York (1959).

[G 5] F.R. Gantmacher, *Theory of Matrices*, Vol II Chelsea, New York (1959).

[G 6] J.D. Gardiner, A.J. Laub, *A generalization of the matrix sign function solution for algebraic Riccati equations*, Int. J. Control 44 (Sep. 1986), 823–832.

[G 7] T. Geerts, *All optimal controls for the singular linear–quadratic problem without stability; a new interpretation of the optimal cost*, Lin. Alg. Appl. 116 (1989), 135–181.

[G 8] T. Geerts, *Structure of linear quadratic Control*, Ph.d Thesis, Technical University of Eidnhoven, Netherlands (1989).

[G 9] T. Geerts, *The Algebraic Riccati Equation and Singular Optimal Control*, Lecture Notes of the Workshop on "The Riccati Equation in ontrol, Systems and Signals", S. Bittanti, Ed., Pitagora Editrice Bologna (June 1989), 415–420.

[G 10] T. Geerts, *Infinite-horizon linear quadratic control: A state of the art*, Sonder-forschungsbereich 343, "Diskrete Strukturen in der Mathematik", Universität Biele-feld, Preprint 90-038 (1990).

[G 11] T. Geerts, V. Mehrmann, *Linear differential equations with constant coefficients: A distributional approach*, Report 90-073, Sonderforschungsbereich 343, "Diskrete Struk-turen in der Mathematik", Universität Bielefeld (1990).

[G 12] G.H. Golub, C.F. Van Loan, *Matrix Computations*, The Johns Hopkins University Press, Baltimore, Md (1983).

[G 13] I. Gohberg, P. Lancaster, L. Rodman, *Matrices and Indefinite Scalar Products*, Birkhäuser, Basel (1983).

[G 14] I. Gohberg, P. Lancaster, L. Rodman, *On hermitian solutions of the symmetric algebraic Riccati equations*, SIAM J. Contr. and Opt. 24 (1986), 1323–1334.

[H 1] S.J. Hammarling, *Some Notes on the use of orthogonal similarity transforms in control*, NPL Report DITC 8/82 (Aug. 1982).

[H 2] S.J. Hammarling, *Numerical solution of the stable, nonnegative definite Lyapunov equation*, IMA J. of Num. Anal. 2 (1982), 303–323.

[H 3] S.J. Hammarling, *Newton's method for solving the algebraic Riccati equation*, NPL Report DITC 12/82 (Sept. 1982).

[H 4] S.J. Hammarling, M.A. Singer, *A canonical form for the algebraic Riccati equation*, Proc. of the MTNS meeting (1983).

[H 5] M.C.J. Hautus, *Controllability and observability conditions of linear autonomous systems*, Proc. Kon. Ned. Akad. Wetensch. Ser A, 72 (1969), 443–448.

[H 6] G.A. Hewer, *An iterative technique for the computation of the steady state gains for the discrete optimal regulator*, IEEE Trans. Autom. Control AC–16 (1971), 382–384.

[H 7] G. Hewer, G. Nazaroff, *A survey of numerical methods for the solution of algebraic Riccati Equations*, Technical Report, Naval Weapons Center, China Lake, Ca. (1974).

[H 8] R.A. Horn, C.R. Johnson, *Matrix Analysis*, Cambridge University Press, Cambridge (1985).

[H 9] A.S. Householder, *The Theory of Matrices in Numerical Analysis*, Dover Publications, New York (original Blaisdall 1964) (1975).

[H 10] J.L. Howland, *The sign matrix and the separation of matrix eigenvalues*, Lin. Alg. Appl. 49 (1983), 221–332.

[H 11] L.K. Hua, *On the theory of automorphic functions of a matrix variable, I, Geometrical basis*, Amer. J. Math. 66 (1944), 470–488.

[H 12] L.K. Hua, *On the theory of automorphic functions of a matrix variable, II, The classification of hypercircles under the symplectic group*, Amer. J. Math. 66 (1944), 531–563.

[H 13] P.C. Hughes, R.E. Skelton, *Controllability and observability of linear matrix second order systems*, ASME J. Applied Mech. 47 (1980), 415–420.

[I 1] F. Incertis, *A new formulation of the algebraic Riccati problem*, IEEE Trans. Autom. Control AC–26 (1981), 768–770.

[J 1] C.G.J. Jacobi, *Über ein leichtes Verfahren, die in der Theorie der Säkularstörungen vorkommenden Gleichungen numerisch aufzulösen*, Crelle's Journal 30 (1846), 51–94.

[J 2] D.H. Jacobson, *Extensions of Linear Quadratic Control*, Optimization and Matrix Theory, Academic Press, New York (1977).

165

[J 3] A. Jennings, J. Halliday, M.J. Cole, *Solution of linear generalized eigenvalue problems containing singular matrices*, J. Inst. Math. Appl. 22 (1978), 401–410.

[K 1] B. Kågström, *RGSVD - An algorithm for computing the Kronecker structure and reducing subspaces of singular A − λB pencils*, SIAM J. Sci. Stat. Comp. 7 (1986), 185–211.

[K 2] W. Kahan, *A survey of error analysis*, Proceedings of the IFIP Congr. Amsterdam (1971), 1214–1239.

[K 3] T. Kailath, *Systems Theory*, Prentice Hall, Englewood Cliffs, N.Y. (1980).

[K 4] R. Kalman, *Contributions to the theory of optimal control*, Boletin Sociedad Matematica Mexicana 5 (1960), 102–119.

[K 5] R.E. Kalman, *Mathematical description of linear dynamical systems*, SIAM J. Control 1 (1963), 182–192.

[K 6] R.E. Kalman, J.E. Bertram, *General synthesis procedure for computer control of single and multiloop linear systems*, Trans. AIEE 77 p. II (1958).

[K 7] R.E. Kalman, P.L. Falb, M.A. Arbib, *Topics in mathematical system theory*, Mac Graw Hill, New York (1969).

[K 8] R.E. Kalman, Y.C. Ho, K.S. Narendra, *Controllability of linear dynamical systems*, Contrib. Diff. Equations 1 (1963), 189–213.

[K 9] C. Kenney, G. Hewer, *The sensitivity of the algebraic and differential Riccati equation*, Scientific Computation Laboratory, University of California of Santa Barbara, Preprint (1990).

[K 10] C. Kenney, A.J. Laub, *Condition estimation for matrix functions*, SIAM J. Matrix Anal. Appl. 10 (1989), 191-209.

[K 11] C. Kenney, A.J. Laub, *Rational iterative methods for the matrix sign function*, Scientific Computation Laboratory, University of California of Santa Barbara, Technical report 89–08 (April 1989).

[K 12] C. Kenney, A.J. Laub, *Newton's Method for polar decomposition and the matrix sign function*, Scientific Computation Laboratory, University of California of Santa Barbara, Technical Report SCL 89–11 (May 1989).

[K 13] C. Kenney, A.J. Laub, *Polar decomposition and matrix sign function condition estimates*, Scientific Computation Laboratory, University of California of Santa Barbara, Technical report 89–02 (June 1989).

[K 14] C. Kenney, A.J. Laub, E.A. Jonckheere, *Positive and negative solutions of dual Riccati equations by matrix sign function iteration*, Scientific Computation Laboratory, University of California of Santa Barbara, Technical Report SCL 89–04 (May 1989).

[K 15] C. Kenney, A.J. Laub, M. Wette, *A stability-enhancing scaling procedure for Schur-Riccati solvers*, Systems and Conrol Letters 12 (1989), 241–250.

[K 16] C. Kenney. A.J. Laub, M. Wette, *Error bounds for Newton refinement of solutions to algebraic Riccati equations*, Scientific Computation Laboratory, University of California of Santa Barbara, Technical Report SCL 88–02 (December 1988).

[K 17] D.L. Kleinman, *On an iterative technique for Riccati equation computations*, IEEE Trans. Autom. Control AC–13 (1968), 114–115.

[K 18] H.W. Knobloch, H. Kwakernaak, *Lineare Kontrolltheorie*, Springer, Berlin (1985).

[K 19] M.M. Konstantinov, P.H. Petkov, N.D. Christov, *Pertubation analysis of the continuous and discrete matrix Riccati equations*, Proc. American Control Conference, Seattle, Wa 1 (1986), 636–639.

[K 20] M.M. Konstantinov, P.H. Petkov, N.D. Christov, *Perturbation Analysis of Linear Control Problems*, Preprint, Institute of Engineering Cybernetics and Robotics, Akad. G. Vonchev Str., Bl, 2, 1113 Sofia, Bulgaria.

[K 21] L. Kronecker, *Algebraische Reduction der Schaaren bilineare Formen*, S.B. Akad. Berlin (1980), 1225–1237.

[K 22] V. Kučera, *The structure and properties of time optimal discrete linear control*, IEEE Trans. Autom. Control AC–16 (1971), 375–377.

[K 23] V. Kučera, *A contribution to matrix quadratic equations*, IEEE Trans. Autom. Control (1972), 344–347.

[K 24] V. Kučera, *A review of the matrix Riccati Equation*, Kybernetica 9 (Prague) (1973).

[K 25] V. Kučera, *Algebraic Riccati equation: Symmetric and definite solutions*, Lecture Notes of the Workshop on "The Riccati Equation in Control, Systems and Signal", S. Bittanti, Ed., Pitagora Editrice Bologna (1989), 73–75.

[K 26] P. Kunkel, V. Mehrmann, *Numerical solution of Riccati differential algebraic equations*, Lin. Alg. Appl. 137/138 (1990), 39–66.

[K 27] P. Kunkel, V. Mehrmann, *Smooth factorizations of matrix valued functions and their derivatives*, Preprint 63, Inst. f. Geometrie und Praktische Mathematik, RWTH Aachen, to appear in Numerische Mathematik (1991).

[K 28] H. Kwakernaak, R. Sivan, *Linear Optimal Control Systems*, Wiley Interscience, New York (1972).

[L 1] P. Lancaster, A.C.M. Ran, L. Rodman, *Hermitian solutions of the discrete algebraic Riccati equation*, Int. J. Control 44 (1986), 777–802.

[L 2] P. Lancaster, A.C.M. Ran, L. Rodman, *An existence and monotonicity theorem for the discrete algebraic matrix Riccati equation*, Lin. and Multil. Algebra 20 (1987), 353–361.

[L 3] P. Lancaster, L. Rodman, *Existence and uniqueness theorems for the algebraic Riccati equation*, Int. J. Control 32 (1980), 285–309.

[L 4] P. Lancaster, L. Rodman, *Solutions of the Continuous and Discrete–time Algebraic Riccati Equations: A Review*, Preprint, submitted to Lecture Notes of the Workshop on "The Riccati Equation in Control, Systems and Signal", S. Bittanti, Ed., Pitagora Editrice Bologna.

[L 5] P. Lancaster, M. Tismenetsky, *The Theory of Matrices*, 2nd edition, Academic Press, Orlando (1985).

[L 6] D.G. Laniotis, *Partitional Riccati solutions and integration free doubling algorithms*, IEEE Trans. Autom. Control AC–21 (1976), 677–689.

[L 7] A.J. Laub, *A Schur method for solving algebraic Riccati equations*, IEEE Trans. Autom. Control AC–24 (1979), 913–921.

[L 8] A.J. Laub, *On computing balancing transformations*, Proceedings JACC, San Francisco, Ca. (1980).

[L 9] A.J. Laub, *Schur techniques in invariant inbedding methods for solving two–point boundary value problems*, Proc. 21st IEEE Conference on decision & control, vol 1, Orlando Florida (1982), 56–61.

[L 10] A.J. Laub, *Schur techniques for Riccati differential equations*, D. Hinrichsen, A. Isidori, Edts. Feedback Control of Linear and Nonlinear Systems, Springer, Berlin (1982), 165–174.

[L 11] A.J. Laub, *Numerical linear algebra aspects of control design computations*, IEEE Trans. Autom. Control AC–30 (1985), 97–108.

[L 12] A.J. Laub, *Algebraic aspects of generalized eigenvalue problems for solving Riccati equations*, Comp. & Comb. Methods in System Theory, Byrnes, Lindquist, Edts Elsevier, North Holland (1986), 213–227.

[L 13] A.J. Laub, *Invariant subspace methods for the numerical solution of algebraic Riccati equations*, Report SCL 90-02, Scientific Computing Laboratory, University of California, Santa Barbara (1990).

[L 14] A.J. Laub, K. Meyer, *Canonical forms for symplectic and Hamiltonian matrices*, Celest. Mechanics 9 (1974), 213–238.

[L 15] C. Lawson, R. Hanson, D. Kincaid, *Basic linear algebra subprograms for FORTRAN usage*, ACM Trans. Math. Softw. 5 (1979), 308–323.

[L 16] K.H. Lee, *Generalized eigenproblem structures and solution methods for Riccati equations*, Ph.D. thesis, Univ. of South Calif. (1983).

[L 17] E.B. Lee, L. Markus, *Foundations of Optimal Control Theory*, John Wiley, New York (1967).

[L 18] W. Leontief, *The dynamic inverse in contributions to input output analysis*, A.P. Carter, A. Brody Edts, North Holland, Amsterdam (1970).

[L 19] W. Levine, M. Athans, *On the optimal error regulation of a string of moving vehicles*, IEEE Trans. Autom. Control AC–11 (1966), 355–361.

[L 20] F.L. Lewis, K. Ozcaldiran, *Reachability and controllability for descriptor systems*, Proc. 27, Midwest Symp. on Circ Syst., Morgantown, WV (1984).

[L 21] W.W. Lin, *A new method for computing the closed loop eigenvalues of a discrete–time algebraic Riccati equation*, Lin. Alg. Appl. 96 (1987), 157–180.

[L 22] W.W. Lin, *On Schur type decompositions for Hamiltonian and symplectic pencils*, Preprint Institute of Applied Mathematics, National Tsing Hua University, Taiwan (1990).

[L 23] A.G. Luenberger, A. Arbel, *Singular Dynamic Leontief systems*, Econometrica 45 (1977), 991–995.

168

[M 1] A. Mac Farlane, *An eigenvector solution of the optimal linear regular problem*, J. Electr. and Control 14 (1965), 643–654.

[M 2] M. Maki, *Numerical Solution of algebraic Riccati equation for single input systems*, IEEE Trans. Autom. Control AC–17 (1972), 264–265.

[M 3] K. Martensson, *On the matrix Riccati equation*, Info. Sci. 3 (1971), 17–49.

[M 4] K. Martensson, *New approaches to the numerical solution of optimal control problems*, Lund. Inst. Techn. Report 7206, Dissertation Lund. Univ. (1972).

[M 5] V. Mehrmann, *Der SR-Algorithmus zur Berechnung der Eigenwerte einer Matrix*, Diplomarbeit Universität Bielefeld (1979).

[M 6] V. Mehrmann, *A symplectic orthogonal method for single input or single output discrete time optimal linear quadratic control problems*, SIAM Jour. Matrix Anal. Appl. (1988), 221–248.

[M 7] V. Mehrmann, *Existence, uniqueness and stability of solutions to singular linear quadratic optimal control problems*, Lin. Alg. Appl. 121 (1989), 291–331.

[M 8] V. Mehrmann, *The Linear Quadratic Control Problem: Theory and Numerical Algorithms*, Habilitationsschrift, Universität Bielefeld (1988).

[M 9] V.Mehrmann, G.M. Krause, *Linear transformations which leave controllable multi inpput descriptor systems controllable*, Lin. Alg. Appl. 120 (1989), 47–64.

[M 10] V. Mehrmann, E. Tan, *Defect correction methods for the solution of algebraic Riccati equations*, IEEE Trans. Autom. Control AC–33 (1988), 695–698.

[M 11] H.B. Meyer, *The matrix equation, $AZ + B - ZCZ - ZD = 0$*, SIAM J. Appl. Math. 30 (1976), 136–142.

[M 12] M.L. Michelson, *On the eigenvalue, eigenvector method for solution of the stationary discrete matrix Riccati equation*, IEEE Trans Autom. Control AC–24 (1979), 480–481.

[M 13] G.B. Moler, G.W. Stewart, *An algorithm for generalized matrix eigenvalue problems*, SIAM J. Num. Anal. 10 (1975), 241–256.

[M 14] B.P. Molinari, *Equivalence relations for the algebraic Riccati equation*, SIAM J. Control 11 (1973), 272–285.

[M 15] B. Molinari, *Algebraic solution of matrix linear equations in control theory*, Proc. Inst. Electr. Engineers, 116 (1969), 1748–1754.

[M 16] B.P. Molinari, *The stabilizing solution of the algebraic Riccati equation*, SIAM J. Control 11 (1973), 262–271.

[M 17] B.P. Molinari, *The time invariant linear quadratic optimal control problem*, Automatica 13 (1977), 347–457.

[M 18] P.C. Müller, *Allgemeine lineare Theorie für Rotorsysteme ohne oder mit kleinen Unsymmetrien*, Ingenieur Archiv 51 (1981), 61–74.

[N 1] R. Nikoukhah, A.S. Willsky, B. Levy, *Kalman filtering and Riccati equations for descriptor systems*, Rapports de Recherche N0. 1186, INRIA, Le Chesnay, France (1990).

[P 1] M. Packer, T.E. Bullock, *Ordering and stability properties of the Riccati equation*, Nat. Res. Inst. for Math. Sci. Dep., WISK 264, Pretoria (1977).

[P 2] C. Paige, C.F. Van Loan, *A Schur decomposition for Hamiltonian matrices*, Lin. Alg. Appl. 41 (1981), 11–32.

[P 3] L. Pandolfi, *Controllability and stabilization for linear systems of algebraic and differential equations*, J. Opt. Th. Appl. 30 (1980), 241–254.

[P 4] L. Pandolfi, *On the regulator problem for linear degenerate Control Systems*, J. Opt. Th. Appl. 33 (1981), 241–254.

[P 5] T. Pappas, A.J. Laub, N.R. Sandell, *On the Numerical solution of the discrete–time algebraic Riccati equation*, IEEE Trans. Autom. Control AC–25 (1980), 631–641.

[P 6] B.N. Parlett, *The Symmetric Eigenvalue Problem*, Prentice Hall, Englewood Cliffs, N.J. (1980).

[P 7] B.N. Parlett, C. Reinsch, *Balancing a matrix for calculation of eigenvalues and eigenvectors*, Numer. Math. 13 (1969), 296–304.

[P 8] R.V. Patel, M. Toda, *On norm bounds for algebraic Riccati and Lyapunov equations*, IEEE Trans. Autom. Control AC–23 (1978), 87–88.

[P 9] H.J. Payne, L.M. Silverman, *On the discrete time algebraic Riccati equation*, IEEE Trans. Autom. Control AC–18 (1973), 226–234.

[P 10] G. Peters, J. Wilkinson, $Ax = \lambda Bx$ *and the generalized eigenproblem*, SIAM J. Num. Anal. 7 (1970), 479–492.

[P 11] I.R. Petersen, *The matrix Riccati equation in state feedback H–infinity control and in the stabilization of uncertain systems with Norm Bounded Uncertainty*, Lecture Notes of the Workshop on "The Riccati Equation in Control, Systems and Signal", S. Bittanti, Ed., Pitagora Editrice Bologna (1989), 51–56.

[P 12] P.H. Petkov, N.D. Christov, M.M. Konstantinov, *A posteriori error analysis of the generalized Schur approach for solving the discrete matrix Riccati equation*, Preprint, Department of Automatics, High School of Meachnical amd Electrical Engineering, 1756 Sofia, Bulgaria.

[P 13] P.H. Petkov, N.D. Christov, M.M. Konstantinov, *Numerical Properties of the generalized Schur approach for solving the discrete mariz Riccati equation*, Proceedings of the 18th Spring Conference of the Union of Bulgarian Mathematicians, Albena (1989).

[P 14] P.H. Petkov, N.D. Christov, M.M. Konstantinov, *On the numerical properties of the Schur approach for solving the matrix Riccati equation*, Systems and Control Letters, 9 (1987), 197–201.

[P 15] L.S. Pontryagin, V. Boltyanskii, R. Gamkrelidze, E. Mishenko, *The Mathematical Theory of Optimal Processes*, Interscience, New York (1962).

[P 16] J.E. Potter, *Matrix quadratic solutions*, SIAM J. Appl. Math. 14 (1966), 496–500.

[R 1] W.F. Ramirez, S. Park, *The Riccati equation in the control of biotechnology processes*, Lecture Notes of the Workshop on "The Riccati Equation in Control, Systems and Signal", S. Bittanti, Ed., Pitagora Editrice Bologna (1989), 24–27.

[R 2] A.C.M. Ran, *Hermitian Solutions of the Discrete Algebraic Riccati Equation*, Lecture Notes of the Workshop on "The Riccati Equation in Control, Systems and Signal", S. Bittanti, Ed., Pitagora Editrice Bologna (1989), 93–94.

[R 3] A.C.M. Ran, L. Rodman, *The algebraic matrix Riccati equation*, Operator Theory: Advances and Applications, vol 12 (1984), 351–381.

[R 4] W.T. Reid, *Riccati Differential equations*, Academic Press, New York (1972).

[R 5] D. Repperger, *A square root of a matrix approach to obtain the solution to a steady state matrix Riccati equation*, IEEE Trans. Autom. Control AC–21 (1976), 786–787.

[R 6] W. Rheinboldt, *On the computation of multi-dimensional manifolds of parametrized equations*, Numer. Math 53 (1988), 165–181.

[R 7] J.R. Rice, *A theory of condition*, SIAM J. Numer. Anal. 3 (1966), 287–310.

[R 8] J.R. Rice, *Matrix Computation and Mathematical Software*, Mac Graw Hill, New York (1981).

[R 9] J. Roberts, *Linear model reduction and solution of the algebraic Riccati equation by the use of the sign function*, Int. J. Control 32 (1980), 667–687.

[R 10] L. Rodman, *On extremal solutions of the algebraic Riccati equation*, in: Lectures in Applied Mathematics, (eds. C.I. Byrnes and C.F. Martin), AMS, Providence, RI 18 (1980), 311–327.

[R 11] H.H. Rosenbrock, *State Space and Multivariable Theory*, Wiley, New York (1970).

[R 12] A. Ruhe, *Algorithms for the nonlinear eigenvalue problem*, SIAM J. Numer. Anal. 10 (1973), 674–698.

[S 1] A.P. Sage, *Optimum Systems Control*, Prentice Hall, Englewood Cliffs, N.J. (1966).

[S 2] N.R. Sandell, *On Newton's method for Riccati equation solution*, IEEE Trans. Autom. Control AC–19 (1974), 254–255.

[S 3] I. Schur, *On the characteristic roots of a linear substitution with an application to the theory of integral equations*, Math. Ann. 66 (1909), 488–510.

[S 4] M.A. Shayman, *Homogeneous indices, feedback invariants and control system structure theorem for generalized linear systems*, SIAM J. Control. Opt. 26 (1988), 387-400.

[S 5] M.A. Shayman, *Pole placement by dynamic compensation for descriptor systems*, Automatica, to appear.

[S 6] M.A. Shayman, Z. Zhou, *Feedback control and classification of generalized linear systems*, IEEE Trans. Autom. Control AC–32 (1987), 483–494.

[S 7] H.A. Shubert, *Analytic solution for the algebraic Riccati equation*, IEEE Trans. Autom. Control AC–19 (1974), 255–256.

[S 8] L.M. Silverman, *Discrete Riccati equations: Alternative algorithms, asympt. properties and system theory interpretations*, in Advances in Control Systems, Leondes Ed., Academic Press, New York (1976).

[S 9] M. Simaan, *A note on the stabilizing solution of the algebraic Riccati equation*, Int. J. Control 20 (1974), 239–241.

[S 10] M.A. Singer, S.J. Hammarling, *The algebraic Riccati equation. A summary review of some available results*, NPL Report, DITC 23/83 (Jan. 1983).

[S 11] B.T. Smith, J.M. Boyle, J.J. Dongarra, B.S. Garbow, Y. Ikebe, V.C. Klema, C.B. Moler, *Matrix Eigensystem Routines-EISPACK Guide*, Lecture Notes in Computer Science 6, Springer, Berlin (1976).

[S 12] V.W. de Spinadel, *On optimal control*, Lin. Alg. and its Role in Syst. Theory, Cont. Math., vol 47, AMS (1984), 111–121.

[S 13] G.W. Stewart, *Error bounds for approximate invariant subspaces of closed linear operators*, SIAM J. Num. Anal. 8 (1971), 796–808.

[S 14] G.W. Stewart, *On the sensitivity of the eigenvalue problem $Ax = \lambda Bx$*, SIAM J. Num. Anal. 9 (1972), 669–686.

[S 15] G.W. Stewart, *Introduction to Matrix Computation*, Academic Press, New York (1973).

[S 16] G.W. Stewart, *Error and perturbation bounds for subspaces associated with certain eigenvalue problems*, SIAM Review 15 (1973), 727–764.

[S 17] G.W. Stewart, *HQR 3 and EXCHN 6 , FORTRAN subroutines for calculating and ordering the eigenvalues of a real upper Hessenberg matrix*, ACM Trans. Math. Softw. 2 (1975), 275–280.

[S 18] G.W. Stewart, *Simultaneous iteration for computing invariant subspaces of non Hermitian matrices*, Numer. Math. 25 (1976), 123–136.

[S 19] G.W. Stewart, *Perturbation Theory for the Generalized Eigenvalue Problem*, Academic Press, New York (1978).

[T 1] S. Tan, A unified approach for analysis and control of multivariable non–causal systems, Phd thesis Catholic University of Leuven, Belgium (1987).

[T 2] R.C. Thompson, *The characteristic polynomial of a principal submatrix of a Hermitian pencil*, Lin. Alg. Appl. 14 (1976), 135–177.

[T 3] J.S. Thorp, *The singular pencil of a linear dynamical system*, Lecture Notes of the Workshop on "The Riccati Equation in Control, Systems and Signal", S. Bittanti, Ed., Pitagora Editrice Bologna (1973), 577–596.

[T 4] H.L. Trentelman, *The totally singular linear quadratic problem with indefinite cost*, Lecture Notes of the Workshop on "The Riccati Equation in Control, Systems and Signal", S. Bittanti, Ed., Pitagora Editrice Bologna (1989), 120–128.

[U 1] V.V. Udilov, *On the optimization of linear control systems with a symmetric main part*, J. Inst. Kibernetik Akad. Nauk. Ukraine, Kiev URSS (1968), 63–71.

[V 1] A.J.J. Van der Weiden, O.H. Bosgra, *The determination of structural properties of a linear system by operations of system similary: 2, Non proper systems in generalized state space form*, Int. J. Control 32 (1980), 489–537.

[V 2] P. Van Dooren, *The computation of Kronecker's canonical form of a singular pencil*, Lin. Alg. Appl. 27 (1979), 103–121.

[V 3] P. Van Dooren, *The generalized eigenstructure problem in linear system theory*, IEEE Trans. Autom. Control AC-26 (1981), 111–129.

[V 4] P. Van Dooren, *A generalized eigenvalue approach for solving Riccati equations*, SIAM J. Sci. Stat. Comp. 2 (1981), 121–135.

[V 5] P. Van Dooren, *Algorithms 590 DSUBSP and EXCHQZ, FORTRAN subroutines for computing deflating subspaces with specified spectrum*, ACM Trans. Math. Softw. 8 (1982), 376–382.

[V 6] P. Van Dooren, M. Verhaegen, *On the use of unitary state-space transformations*, Lin. Alg. and its Role in System. Theory, Contemp. Math. 47, AMS (1984), 447–465.

[V 7] C.F. Van Loan, *A symplectic method for approximating all the eigenvalues of a Hamiltonian matrix*, Lin. Alg. Appl. 61 (1984), 233–251.

[V 8] D.R. Vaughan, *A negative exponential solution for the matrix Riccati equation*, IEEE Trans. Autom. Control AC–14 (1969), 72–75.

[V 9] D.R. Vaughan, *A nonrecursive algebraic solution for the discrete Riccati equation*, IEEE Trans. Autom. Control AC–15 (1970), 597–599.

[V 10] G.C. Verghese, B.C. Lévy, T. Kailath, *A general state space for singular systems*, IEEE Trans. Autom. Control AC–26 (1981), 811–831.

[V 11] G.C. Verghese, P. Van Dooren, T. Kailath, *Properties of the system matrix of a generalized state space system*, Int. J. Control 30 (1979), 235–243.

[W 1] R.A. Walker, A. Emami–Naeini, P. Van Dooren, *A generalized algorithm for solving the algebraic Riccati equation*, Proc. 21st, IEEE Conf. Dec & Control (1982), 68–72.

[W 2] R.C. Ward, *The combination shift QZ–algorithm*, SIAM J. Num. Anal. 12 (1975), 835–853.

[W 3] R.C. Ward, *Balancing the generalized eigenvalue problem*, SIAM J. Sci. Stat. Comp. 2 (1981), 141–152.

[W 4] D.S. Watkins, *On GR Algorithms for the Eigenvalue Problem*, Proceedings of NATO Advanced Study Institute on Numerical Linear Algebra, Digital Signal Processing, and Parallel Algorithms, Leuven, Belgium (1988).

[W 5] D.S. Watkins, L. Elsner, *Chasing algorithms for the eigenvalue problem*, Preprint, Washington State University, Pullman, WA (1989).

[W 6] D.S. Watkins, L. Elsner, *Self-similar flows*, Lin. Alg. Appl. 110 (1988), 213–242.

[W 7] J.H. Wilkinson, *Rounding Errors in Algebraic Processes*, Prentice Hall, Englewood Cliffs, New Jersey (1963).

[W 8] J.H. Wilkinson, *The Algebraic Eigenvalue Problem*, Claredon Press, Oxford (1965).

[W 9] J.H. Wilkinson, *Kronecker's canonical form and the QZ algorithm*, Lin. Alg. Appl. 28 (1979), 285–303.

[W 10] J.H. Wilkinson, C. Reinsch, *Linear Algebra*, Handbook for Automatic Computation, Springer, Berlin (1971).

[W 11] J.C. Willems, *Least squares stationary optimal control and the algebraic Riccati equation*, IEEE Trans. Autom. Control AC–16 (1971), 621–634.

[W 12] J.C. Willems, *On the existence of a nonpositive solution to the Riccati equation*, IEEE Trans. Aut. Control AC–19 (1974), 592–593.

[W 13] J.C. Willems, *The Linear Matrix Inequality*, Lecture Notes of the Workshop on "The Riccati Equation in Control, Systems and Signal", S. Bittanti, Ed., Pitagora Editrice Bologna (1989), 46–47.

[W 14] J.C. Willems, A. Kitapci, L. Silverman, *Singular optimal control: A geometric approach*, SIAM J. Control Opt. 24 (1986), 323–337.

[W 15] H.K. Wimmer, *The algebraic Riccati equation without complete controllability*, SIAM J. Alg. Discr. Methods 3 (1982), 1–12.

[W 16] H.K. Wimmer, *Monotonicity of maximal solutions of algebraic Riccati equations*, Systems and Control Letters 5 (1985), 317–319.

[W 17] H.K. Wimmer, *Stabilizing and unmixed solutions of the discrete–time algebraic Riccati equation*, Proceedings of the Workshop on "The Riccati Equation in Control, Systems and Signal", S. Bittanti, Ed., Pitagora Editrice Bologna (1989), 95–98.

[W 18] W.M. Wonham, *On a matrix equation of stochastic control*, SIAM J. Control Opt. 6 (1968), 681–697.

[W 19] W.M. Wonham, *Linear Multivariable Control*, Lecture Notes in Economics and Math. Systems 101, Springer, Berlin (1974).

[Y 1] E.L. Yip, R.F. Sincovec, *Solvability controllability and observability of continous descriptor systems*, IEEE Trans. Autom. Control AC–26 (1981), 702–707.

[17] J.C. Willems, On the variance of expo... ... IEEE Trans. Aut. Control AC-25 (1980), 291-305.

[18] J.C. Willems, The Riccati equation in ... ... "Three Decades in Control Systems and Signal ...", Blanco, Bologna, ...

[19] W.M. Wonham, H.J. Kushner, Stochastic ... ... SIAM J. Control ... 2 (1968), 347-371.

[20] W.M. Wonham, On a matrix Riccati equation ... ... SIAM J. Control ... 6 (1968), 312-326.

[21] W.M. Wonham, On the separation theorem of stochastic control, SIAM J. Control 6 (1968), 312-326.

[22] ... ...

[23] ... ...

[24] ... ... IEEE Trans. ... ...